CHAINS OF OPPORTUNITY

Ohio History and Culture

Mark D. Bowles

CHAINS OF OPPORTUNITY

THE UNIVERSITY OF AKRON AND THE EMERGENCE OF THE POLYMER AGE

1909–2007

The University of Akron Press
Akron, Ohio

First Edition 2008
Manufactured in the United States of America.
All inquiries and permission requests should be addressed to the Publisher,
The University of Akron Press, Akron, Ohio 44325-1703.
www.uakron.edu/uapress

12 11 10 09 08 5 4 3 2 1

Library of Congress Cataloging-in-Publication Data

Bowles, Mark D.
Chains of opportunity : The University of Akron and the emergence of the polymer age, 1909–2007 / Mark D. Bowles.
p. cm.
ISBN 978–1–931968–53–9 (cloth : alk. paper)
1. University of Akron. College of Polymer Science and Polymer Engineering—History. 2. Polymers—Study and teaching (Higher)—Ohio—Akron—History. I. University of Akron. II. Title.
QD47.5.O45U553 2008
668.907'077136—dc22

2007044709

The paper used in this publication meets the minimum requirements of American National Standard for Information Sciences—Permanence of Paper for Printed Library Materials, ansi z39.48–1984. ∞

TABLE OF CONTENTS

LIST OF ILLUSTRATIONS

In memory of my father
Donald L. Bowles

PREFACE

As a child in the 1970s, I seemed to grow up at the University of Akron. My father was an administrator there for over thirty years and on Saturdays he would take me with him to his office and share stories of its past. While he worked, I would sometimes play on the winding double-staircase of Buchtel Hall, dropping my toy astronauts over the railing, thinking the place was immensely austere and breathtaking. At the dinner table each night, he invoked iconic names in the school's history, like Auburn, Morton, Sumner, and Kelley, and recounted tales of their exploits. This was an exciting era of construction at the university, and I can remember making a wager with my father over what building would be completed first, E.J. Thomas Performing Arts Hall or Bierce Library. I lost the bet but realized later that he probably had some inside information since he worked in the facilities planning office. My father's love of the university was infectious, and there was never a question of what I was going to do after high school. Instead, the question was what major I would someday choose at the University of Akron. I picked history and earned my first graduate degree there.

My father concluded his career at the university in 1990 as the first associate dean of the College of Polymer Science and Polymer Engineering. Over the years, "polymers" had become an almost magical word in our household. They were the material of the future, a worthy academic pursuit, and a potential savior for the Akron region. Sadly, my father was diagnosed with cancer soon after his retirement. In 2001, as I sat with him during his final few days in the hospital, the

Goodyear Polymer Center stood like a guardian in the distance. In the emotional silence of the room, my faraway gaze often focused on the building as I recalled my father's strong voice and his active life. At one point I whispered to him, "You know, dad, you have the best room in the hospital because you can see the polymer building from here." He turned to me slightly and smiled. It was the last time I would ever see my father's smile.

Today, on the east side of the Goodyear Polymer Center, lies a place that pays tribute to my father's life—the Donald L. Bowles Memorial Garden. It sits on what was once Buchtel Avenue in the pedestrian walkway he helped to create, and in the shadow of the building that was so dear to him. The building is a symbol of so many things—polymer excellence, scholarly distinction, civic renaissance—but for me it will always be a symbol of my father. Whenever I see the building on the Akron skyline, especially when it is glowing at night, I can see my father in his tie and hardhat, carrying a tube of blueprints and engaging a group of people about the exciting future that lay ahead. When Dean Frank Kelley contacted me in June 2004 and asked me to write a book recounting the history of the rubber chemistry and polymer programs at the university, I eagerly accepted the challenge. The project was an irresistible opportunity to contribute my voice and perspective to this special place and, in a personal sense, reconnect with the memory of my father.

Books like these are never the product of just one person. There are numerous people who contributed a great deal to this project, and I would like to thank them. First and foremost is Dean Frank Kelley, who invited me into his college and asked me to tell its story to the world. He has been an inspiring leader at the university, and he has played a significant role in establishing its international reputation today. Current University of Akron President Luis M. Proenza has also been a visionary leader who has successfully transformed the campus, physically and aesthetically. I thank him for his participation in this book. Former President William Muse played a critical role in working with Dean Kelley to establish the College of Polymer Science and Polymer Engineering and in building the Goodyear Polymer Center. I deeply appreciate the time and materials he shared with me to support this project, as well as his friendship with my father over the years. It was also a pleasure talking to Priscilla Simmons about her father Hezzleton Simmons, who was president of the university from 1933 to 1951.

Many other people were interviewed for this book and for an earlier DVD project that commemorated the fiftieth anniversary of the first Ph.D. program in polymer chemistry at the University of Akron. These individuals included Stephen Cheng, James D'Ianni, Mark D. Foster, Alan Gent, Lloyd Goettler, H. James Harwood, Sadhan Jana, Joseph Kennedy, Charles Rader, and James White. Those who provided on-camera laboratory interviews at the Goodyear Polymer Center included: William J. Brittain, Gustavo A. Carri, Ali Dhinojwala, Mark D. Foster, Crittenden Ohlemacher, Coleen Pugh, Judit Puskas, Darrell H. Reneker, Alexei P. Sokolov, and Bojie Wang. Laboratory interviews at the Sidney L. Olson Research Center included Mukerrem Cakmak, Kevin Cavicchi, Lloyd Goettler, and Mark D. Soucek. Tayba L. Tahir spoke on the Akron Polymer Training Center. Charles Wilkes shared his thoughts on EPIC with me. The Rubber Division of the American Chemical Society played an important role in supporting this book, and I thank Ed Miller and the Rubber Division for its contributions. Also instrumental from the Akron Global Polymer Academy were Jason Miller and Justin Molenaur, who coordinated and captured many of the interviews on digital media.

Historians always rely heavily upon the skills and resources of archivists and librarians. John V. Miller and Craig A. Holbert at the University of Akron Archives, which houses an excellent collection of university documents, always provided a welcome place for research. Christopher J. Laursen, manager of the John H. Gifford Memorial Library, provided a great deal of assistance for my research. He is a valuable asset to the Rubber Division. I also thank Frank Bove, a reference librarian at the University of Akron's Bierce Library. Rosemary Kolton, senior administrative assistant to Dean Kelley, answered my numerous questions over the course of this project. Kent Marsden, assistant to the dean, assured me that I had all the tools I needed for my research at my disposal. Melissa Bowman and Sarah Thorley assisted with gathering of alumni data. Lynn Patterson transcribed the interviews conducted for this book, and as always I appreciate her professionalism and friendship.

I would also like to extend my gratitude to the University of Akron Press—especially interim director Elton Glaser, production coordinator Amy Freels, business coordinator Carol Slatter, and copy editor Elizabeth Armstrong—for enhancing this manuscript and guiding it through the publication process.

This book was written at a time of transition for the College of Polymer Science and Polymer Engineering. Dean Frank Kelley stepped down in December 2006 after nearly thirty years of service. I would like to thank Stephen Cheng, who was selected as Kelley's successor in August 2007, for sharing his thoughts with me about the new challenges that awaited him.

It is often said that everyone who lives in Akron has a relative who has worked in the rubber industry. My family is no exception. My grandfather, Charles Putman, began a forty-year career at B.F. Goodrich in 1928. My father had close ties to the polymer program at the University of Akron and devoted his life to the school. John Putman, his wife Kay, and son Matthew (my uncle, aunt, and cousin) established Tech Pro, a manufacturer of rubber-testing instruments and software, in Akron in 1983. I thank them for allowing me the opportunity to serve in Tech Pro's administration for the past five years. My mother, Joan, has cared for all of us throughout our lives and supported us in all of our endeavors, all the while managing to keep her hands free of carbon black.

Finally, none of my historical musings would be possible without the love of my wife, Nancy. One of our daughter Isabelle's favorite places to picnic is what she calls "Grandpa Don's Garden" located outside the Goodyear Polymer Center. There we often sit amidst the young birch trees, talk about the past, watch the clouds reflected in the mirrored building, and admire the new Chihuly sculpture at the heart of Frank Kelley Plaza. Those moments of quiet reflection are essential in life, and it is my sincere hope that this book inspires the reader to take time to do the same and reflect upon Akron's rubber and polymer heritage. It is an important part of who we all are, for those of us who still live with its legacy.

INTRODUCTION

Archaeologists have defined past civilizations by the materials that predominated in their societies. The evolution of the use of these materials—stone, bronze, and iron—enabled significant and concurrent changes in the cultures that mastered them. The ability to develop, shape, and use these substances has been one of the measures of a civilization's capabilities, so much so that we now identify them as the Stone Age, Bronze Age, and Iron Age. Historians have also associated the comings and goings of other, more recent, periods with the prevalence of particular types of materials. Brooke Hindle identified America's pre-Civil War period as its "Wooden Age," and he characterized the society as one that was "pervasively conditioned by wood."[1] This dependence slowly subsided after 1850 and was replaced in part by metals and the corresponding products of industrialization.

There was one common characteristic that defined these former ages. The materials in question all came from nature. However, at the dawning of the twentieth century, scientists and engineers began to master a new class of materials called polymers. Although some of these occurred in nature (like spider silk or natural rubber), other polymers were made in laboratories and factories and had properties superior to those existing in the natural world. Over the past century, the ability to create and manipulate natural and synthetic polymers has resulted in a revolutionary transformation of our own material culture. One hundred years ago, the artifacts of everyday life were made of wood, stone, glass, or metal. This is no longer the case, as polymers now often replace these substances and enhance the quality of the products made with them. The result is that polymers have become one of the most significant scientific and engineering achievements in the past century. The father of polymer education, Herman Mark, suggested in 1982 that society was "coming to an age in polymer science."[2] In 1976, an author in *Chemical Week* simply christened this the "polymer age."[3]

Although it is often difficult and arbitrary to pinpoint dates of when "ages" like this begin or end, it might be argued that the Polymer Age matured in the late 1970s. While the realization that polymers were macromolecules took place in the 1920s, it was not until a half century later that the consumption of synthetic polymeric materials overtook metals in the United States.[4] In 1973, the United States produced the same amount of synthetic polymers as it did steel—15.5 million cubic meters.[5] In 1976, the United States consumed more plastics than metal on a per volume basis for the first time, and plastics became the most used synthetic material in the country.[6] The rapidity with which polymers became so prevalent is an astonishing story. In 1900, only a handful of chemists concerned themselves with materials that would later be known as "polymers," and none knew what they truly were. In the 1930s, the United States produced virtually no synthetic polymers. Forty years later, it produced twenty-three million metric tons. Statistics like these were strong indicators that a new material age had begun. Today half of all chemists work with them.[7]

Cultural indicators also point to a dramatic change since the 1970s in how we value polymeric products (plastics, rubber, and textiles). Consider how plastics were viewed by Hollywood and pop music stars in the 1960s. In the 1967 film *The Graduate*, Benjamin Braddock (played by Dustin Hoffman) was told by one of his father's colleagues that the future lay in one word—"plastics." To Braddock, this suggestion conveyed a sense of conformity, selling-out, boredom, and cheap imitation. The audience of the 1960s easily understood how the mere suggestion of "plastics" symbolized the dull life Braddock would eagerly try to escape. It reminded viewers of a similar song released that same year by pop duo Sonny and Cher, who sang about the "plastic man." Their lyrics warned about a man not to be trusted, like someone selling cars on television. He was working day and night, not disclosing he was recruiting everyone for his commercial desires. They sang, "He's doing everyone he can . . . Plastic Man." The Beatles *Rubber Soul* album, released in the mid–1960s, further exemplified the negative connotation of another polymer.

But, an important cultural shift has occurred since then. By the 1970s, the image of plastics began to lose its substandard connotations and instead conveyed the belief that polymers represented a new and potentially better way of life. Symbolizing this phenomenon was an unlikely musician who recognized polymers as a new, dominant material. From 1976 to 1979, legendary punk rock singer Marianne Elliot-Said adopted the pseudonym "Poly Styrene" to make a statement about the significance of the changes she saw around her.[8] She created a character who idealized the glamour of plastic and echoed a youth movement that would eventually come to embrace a new polymer world. She chose the name Poly Styrene for this persona as she herself became infatuated with "plastic" society. One of her songs, a top-thirty hit in London, was called "The Day the World Turned Day-Glo"; in its lyrics she observed that polymers were taking over society. She sang:

> I clambered over mounds and mounds of polystyrene foam
> Then fell into a swimming pool filled with fairy snow
>
> And watched the world turn Day-Glo
> You know you know the world turned Day-Glo you know
>
> I wrenched the nylon curtains back as far as they would go
> Then peered through Perspex window panes at the acrylic road
>
> I drove my polypropylene car on wheels of sponge
> Then pulled into a wimpy bar to have a rubber bun
>
> The x-rays were penetrating thru latex breeze
> Synthetic fibre see-thru leaves fell from the rayon trees

The song was a virtual encyclopedia of the new material world she encountered: polystyrene, Day-Glo, nylon, Perspex, acrylic, polypropylene, latex, and rayon. Polystyrene, known by many trade names, e.g., Styrofoam, is a plastic used for a variety of products, including utensils, CD jewel cases, and models. Day-Glo was created in 1946 after the discovery of new polymer dyes and resins that produced unprecedented bright colors for the plastics, paints, coatings, and textile industry. Nylon was one of the first synthetic polymers, produced in 1935 by Wallace Carothers and used soon thereafter for such products as toothbrush bristles and women's stockings. Perspex is the trade name for a type of acrylic polymer, also called Lucite or Plexiglas. Polypropylene is another thermoplastic polymer used in numerous products. Latices are natural and synthetic polymers disposed in water. These are used in the manufacture of clothing and paint. Rayon is derived from cellulose and was first used as an artificial silk in the 1920s. Metaphorically, as Marianne Elliot-Said (or Poly Styrene) observed, the world did turn Day-Glo (more than just the color: it became polymeric) and remains so today.[9]

This song paralleled the dawn of a new material era—the Polymer Age. The extent of the ubiquity of polymers is best understood by taking a look at a typical day in our own lives, with attention to how often we both knowingly and unknowingly encounter polymers. In 2006, Ed Noga, the editor of *Rubber & Plastics News*, did just that and observed the various ways in which he touched polymers over the course of an average day.[10] He awoke in the morning by pressing the thermoplastic elastomer snooze button on his alarm clock. He had to press it a number of times because his latex foam mattress was too comfortable to leave, without a fight. While getting ready for his day, he encountered numerous other polymer materials well before he even left his house: toothbrushes, combs, razor handles, rubber mats, and shower enclosures. He put on his rubber-soled shoes and clothes made with synthetic fibers, while also likely standing on polymer fiber carpeting. He prepared a breakfast with food that had been stored in rubber-sealed containers. It was a rainy day, and he made sure his children had rubber jackets and boots to keep them dry. He then drove himself to work in a vehicle with polymer components too numerous to mention—hoses, belts, o-rings, gaskets, and tires. These were made of EPDM, natural rubber, silicone, or fluoroelastomers. As he drove, he noticed a road crew applying rubberized asphalt, while another used a polymer paint for the yellow striping down the center of the road. He continued over a bridge with rubber expansion joints and then almost unknowingly drove over a smooth rubber railroad crossing. Once at work, he used his pencil's rubber eraser to eliminate mistakes, made a transatlantic telephone call over fiber-optic wires insulated with ethylene-propylene rubber, and rode an elevator whose vibrations were dampened by rubber. Of course, his mouse, keyboard, and computer were all encased in plastic. After work, he picked up his daughter from soccer practice and lost count of the essential polymer products that she used: rubber balls, plastic shin guards, foam-rubber padding, and rubber-cleated spikes on her shoes.

Noga's experience is typical. Our interactions with polymers are overwhelming and increasing by the day. Polymers now unquestionably dominate our material culture on earth. They have also shaped one of the central images of our brief presence on the moon, as Neil Armstrong's famous "one small step" was made with the imprint of his rubber-soled spacesuit boot. And polymers have become an enthusiastic subject of study for students of all ages, including children creating numerous science fair projects and university-level polymer science and engineering graduates preparing for careers in academia and industry.[11] Graduates of today eagerly seek out the same plastic, polymer industry that Dustin Hoffman's character found to be so hollow and uninspiring.

Ironically, although polymers are now prevalent in our lives, few people can actually state with any accuracy what they are.[12] The word polymer derives from the Greek words *poly* meaning "many" and *mer* meaning "parts." The "parts" are small molecules (monomers), "many" of which can combine to form a very long molecule (polymer). This process is called polymerization. The polymerization process is their most important feature and is analogous to a long necklace (polymer) made from many thousands of beads (monomers). Admittedly, even with this simple definition, the concept of a polymer may still be hard to grasp.

One way to understand polymers is to compare them to metals. Though it might be difficult for the average person to provide a dictionary definition of a metal (a class of elementary substances, crystalline when solid and typically characterized by ductility, opacity, conductivity, and a unique luster when freshly fractured), everybody knows one when they see one. Much as the term "metal" describes a range of materials—gold, silver, or copper—so, too, does the term "polymer." Plastics, resins, rubbers, coatings, and adhesives are commonly made from synthetic polymers. There are also natural polymers such as rubber from trees, silk, spider webs, and skin. Even DNA, the molecule that carries the genetic code, is a polymer.

Another way to understand polymers is to see how their properties change as molecules become longer. For example, the element carbon can take the form of any of the three states of matter (solid, liquid, or gas) simply by changing the length of its polymer chain.[13] When a carbon atom is by itself, along with its bonds to four hydrogen atoms, it is a gas—methane. Increasing the length of this carbon chain to eight atoms and it becomes a liquid—gasoline or octane. Increasing the polymer chain to twenty carbon atoms and the result is a solid—candle wax or eicosane.

It is through the polymerization process that these materials have become so vital to our lives. By arranging monomers (anywhere from eight hundred to one hundred thousand of them at a time) together in many different ways, scientists and engineers are making new polymeric materials to replace wood, stone, glass, and metal. Polymers continue to replace traditional materials because they are often less expensive, more durable, and lighter than their natural counterparts.

While it is now nearly impossible to go about a normal day without encountering polymers, they have ironically often become an unseen substance. Steve Love, Akron journalist and coauthor of *Wheels of Fortune*, wrote, "Polymers have an identity problem. They are, comparatively speaking, invisible. . . ."[14] A characteristic of many great scientific and technological discoveries is that, after they have moved into the mainstream, they transition from sublime wonder to an almost forgotten realm. The electric light bulb is an obvious example—an invention that inspired awe in the nineteenth century, today it is simply an assumed part of our lives.[15]

The history of polymers has this character. After the process of vulcanization was established in 1839, natural rubber quickly became a vital material. The dependence upon it became so great during World War II, that the United States realized if it were cut off from natural rubber sources, its military operation would grind to a halt. Without rubber, the greatest military nation in the world would be quickly rendered powerless. The development of synthetic rubber during World War II led in part to the great variety of the synthetic polymeric materials that surround us now almost invisibly. We are living in the midst of the Polymer Age, and few realize, or even know what this means. As Jeffrey L. Meikle argued in his book *American Plastic*, "most people ignored . . . that the United States entered the 'Plastics Age' in 1979 when the annual volume of plastic exceeded that of steel."[16] We can no longer ignore its presence in our lives, and its history requires telling.

A good starting point for understanding the historical development of polymers is Yasu Furukawa's *Inventing Polymer Science*, which traces the new chemistry created by the efforts of Hermann Staudinger and Wallace Carothers in the 1920s. Ironically, despite a long history, many chemistry textbooks and university programs have failed to address polymers. One observer noted that there was "academic apathy" regarding polymer science through the early 1970s.[17] In 1977, Paul Flory, often called the father of polymer science, said that he was never allowed to teach an undergraduate course in macromolecules at Stanford during the 1960s and 1970s. He argued that the result was that few students in the United States had any broad background in polymers. This remained the case as late as 1993 when Herman Mark, another polymer pioneer, said, "It is difficult to understand why, although half of professional chemists work with polymers, elementary textbooks on organic and physical chemistry hardly mention their existence."[18] Popular newspapers also frequently fail to report the latest polymer advances. Since April 1987, the word "polymer" has only appeared in *USA Today* 156 times. Likewise, since 1851 the *New York Times* has printed the term "polymer science" thirty-six times, and "polymer engineer" only five.

Compared to the number of scholarly studies on other scientific and technological developments like computers, automobiles, and electricity, historians have also been surprisingly quiet on the subject of polymers. The retrospective studies that do exist were largely written by former polymer practitioners. These "insider" stories include memoirs, review articles, and autobiographies. The American Chemical Society and the efforts of Raymond S. Seymour have also produced numerous works that contain much technical detail and scientific perspective. However, historians are beginning to take notice of the field, though biography still dominates their approach. Examples include Frank M. McMillan's *The Chain Straighteners* and Peter J.T. Morris's *Polymer Pioneers*.[19] Historical analysis of the development of rubber have received more attention than the broader topic of polymers. Among these, John Loadman's *Tears of the Tree* and Charles Slack's *Noble Obsession* are popular exceptions.[20]

There are many areas of polymer history that need attention. Furukawa noted, "there are few analytical histories that deal with both the intellectual and social setting of macromolecular chemistry in its formative period."[21] There are still fewer histories that deal with its institutional context. David A. Hounshell and John K. Smith, in their *Science and Corporate Strategy*, have presented an excellent account of the history of the DuPont company, including an important discussion of its significant contributions to polymer research.[22] Other histories of the rubber industry are starting to trace the important role they have played in the evolution of the Polymer Age. One of the best includes Mansel G. Blackford and K. Austin Kerr's *BFGoodrich*.[23] Journalists Steve Love and David Giffels have also told the memorable story of the symbiotic relationship between the city of Akron and the emergence of the rubber industry in their compelling *Wheels of Fortune*.[24] But for the most part, biographical treatments have far superseded the institutional perspective of the history of polymers.

There is no question that academic institutions have been essential to the emergence of the Polymer Age, and their historical neglect is an oversight that requires attention. An institutional perspective can help us understand the intellectual and social foundations from which polymer science and engineering developed. Much as an individual's biography yields important insights into a particular historical context, so, too, does an institutional focus reveal key perspectives. The Polymer Age arose through an international network of researchers working within large academic, industrial, and governmental organizations that developed the scientific and engineering skills to create and commercialize polymers. Among these institutions are academic programs (including the University of Akron, the University of Massachusetts-Amherst, Case Western Reserve University, California Institute of Technology, the University of Southern Mississippi, Virginia Tech, and Brooklyn Polytechnic), industrial organizations (e.g., IG Farben, DuPont, Dow Chemical, Goodyear, Bridgestone/Firestone, B.F. Goodrich, and General Tire), and various government agencies and scientific societies like the Rubber Division of the American Chemical Society.

It is beyond the scope of this book to present a sweeping account of the entire network of academic institutions responsible for the emergence of the Polymer Age over the past century. Instead, this book examines the story within the framework of one academic institution that has played the longest and arguably the most influential role in creating and shaping the Polymer Age—the University of Akron. This university has played a signal role in the emergence of the Polymer Age, starting in 1909 when Charles Knight offered the world's first course in rubber chemistry, marking the birth of a new academic discipline. This course has now evolved into an entire college—the College of Polymer Science and Polymer Engineering—which has been ranked for the last ten years as one of the top two academic institutions in the United States in the field. An examination of the near century in between the foundation of the course and the flourishing of the college demonstrates how long polymer chains became chains of opportunity for an academic discipline, an economic region, and a university.

1

THE BIRTH OF RUBBER CHEMISTRY

1909–1941

CENTURIES AGO, RUBBER—OR "CAOUTCHOUC" AS IT WAS then known—was simply a curious, sappy, milky-white substance that flowed from over three hundred different types of trees and other plants. The *Hevea brasiliensis* was the most well known, and Europeans discovered these sixty-foot-tall "weeping" trees in Central and South America in the fifteenth century. They experimented with the unusual elastic substance they produced, using it first as a bouncing ball. Then, in the 1760s, French researcher Francois Fresnau discovered that it was possible to dissolve natural rubber with turpentine, so that it could be spread upon clothing or other substances. In the 1770s, Joseph Priestley coined the term "rubber" after using it to rub out his pencil marks.[1] By the early nineteenth century, the thriving rubber industry emerged. It produced rubberized clothing and boots that were water-resistant. However, these early rubber products had critical flaws—they melted in the summer and cracked in the winter. In 1839, after Charles Goodyear in America, and Thomas Hancock in Britain, unlocked the vulcanization process through the use of sulfur and heat, the true potential of rubber was unleashed.[2]

In 1855, Charles Goodyear published a book on the vast number of applications and uses for this vulcanized "gum-elastic."[3] He envisioned a new rubberized world in which this miraculous substance would transform virtually every area of life. He foresaw mechanical, electrical, military, and medical uses. Homes would be filled with rubber products. Toys and other recreational items would be made with it. Transportation in stagecoaches and carriages would be more pleasant. Goodyear thought people would begin wearing all manner of rubberized goods, and he even imagined "fancy and ornamental uses." While commercial rubber products had existed before, vulcanization enabled rubber goods to become a viable, large-scale industry. Few other major commodities have experienced such a dramatic change in supply, demand, and cultural significance.[4]

Goodyear's prognostications came true, and the vulcanization of rubber is regarded today as one of the most important industrial developments of the nineteenth century. Rubber's properties were nothing short of revolutionary: it was elastic, waterproof, airtight, moldable, and spreadable. It became the "great shock absorber of the industrial age," with rubber springs cushioning carriages, rubber belts enabling machine gears to turn smoothly, and rubber seals preventing leaks and blocking air. Its resistance to electricity made it an enabling partner in the electrification of everyday life.[5] Rubber tires helped establish the bicycle craze of the 1890s and the corresponding association between freedom and personal transportation. At the turn of the twentieth century, the "automobile age" was also well under way, with some correctly predicting—in 1900—that there would soon be more cars than horses in American cities.[6] Without rubber tires, hoses, and gaskets, these automobiles would have been nothing more than immobile piles of metal and glass. But, despite the centrality of rubber to culture and industry by the start of the twentieth century, there was one unexpected and significant place in which the value of it was virtually ignored—the university.

In 1915, in the *India Rubber Journal*, one chemistry professor lamented, "It is surprising to note that little consideration is given to the chemistry of rubber by our colleges and universities. Organic chemistry textbooks barely make mention of rubber and the inorganic treatises seldom speak of the value of mineral pigments and fillers to the rubber industry."[7] This situation was little changed a decade later. A report by the National Industrial Conference Board in 1925 noted, "There are many undeveloped contacts between the rubber industry and the colleges of the country."[8]

But, there was one institution that stood out above any other in elevating rubber chemistry to academic prominence. In 1909, Charles Knight, a chemistry professor at Buchtel College in Akron, Ohio, offered the first college course that focused specifically upon rubber (in 1913, this institution changed its name to the University of Akron.) In 1917, the journal *Rubber Age* reported that this was the "only educational institution in the world to give instruction in this subject."[9] Knight taught in the perfect place for such an endeavor. His campus sat on a hill, essentially overlooking the most important assemblage of rubber companies ever constructed, including the names Goodrich, Goodyear, and Firestone.

Knight retired just three years after he first taught this course, and he passed on responsibility for its continuation to Hezzleton Simmons, his former laboratory assistant. Simmons quickly established his own reputation and, by the end of his career, became known as "the outstanding American educator of his time in the field of rubber chemistry."[10] He taught a rubber chemistry course for twenty years (from 1913 to 1933) and worked tirelessly to expand it, popularize it, and spread its status internationally. The course, championed and promoted by Knight and Simmons, helped to provide the stimulus for research in rubber, to overcome the rampant secrecy within the rubber industry, and also to educate the next generation of scientists and industrial managers. Everyone, from the rank and file technical workers in the rubber industry to its corporate leaders, eagerly signed up for the course. This pioneering academic course in 1909 remained a nearly one-of-a-kind educational experience until the start of World War II.

THE SECRET WORLD OF RUBBER RESEARCH

In July 1904, the *New York Herald* published an article discussing the emergence of Akron, Ohio, as an industrial leader in the development of rubber goods. With more than twenty established rubber factories, as well as the headquarters for the world's largest rubber corporations, Akron soon became known as the "center of the rubber manufacturing industry of the world," producing over eighty thousand finished tires a day.[11] But it was not immediately obvious to outsiders why Akron, as opposed to other industrial centers in the United States, achieved this prominence. After pondering what it considered a curious phenomenon, the *New York Herald* writer concluded, "Many people wonder what natural advantages Akron offers as a home for rubber factories, and there can be but one answer—there are none."[12]

Despite the opinions of some in competing cities, Akron did offer a number of distinct advantages. In the 1870s, it promoted itself as a city "at the beginning of greatness" and offered incentives to manufacturers looking for a place to start new businesses.[13] The great rubber barons, like Benjamin Franklin Goodrich, who established his roots in Akron in 1870, discovered a rapidly growing city with railroad connections, abundant coal resources, financial supporters, and a low tax rate. Other industrialists soon followed and established thriving organizations: the Miller Rubber Company (1892), Goodyear Tire and Rubber (1898), Kelly-Springfield (1900), Firestone Tire and Rubber Company (1905), and General Tire and Rubber Company (1916). Because of these growing companies, a knowledgeable labor force skilled in the rubber trades emerged. By 1919, over half of all the rubber workers in the United States lived in Ohio, and Akron was becoming one of the fastest growing cities in the nation.[14] The prospects of continued expansion seemed certain.

However, some very important problems emerged. As the rubber industry became more scientifically and technically complex, the training provided on the shop floor was no longer adequate to advance the field.

According to one Goodyear manager in 1912, the rubber and tire manufacturers were "reaching out in every direction for experts—for trained men."[15] By 1917, the editor of the journal *Rubber Age* wrote, "The great handicap on the rubber industry has been the lack of a sufficient number of capable technical men in the rapidly enlarging factories of the country."[16] The university was the primary place for finding these experts, and while many colleges were producing excellent chemistry graduates, none provided specific training in rubber. Rubber chemistry as a university specialty simply did not exist prior to 1909.

Another problem was the secrecy that was ingrained within the rubber industries' corporate culture. This extended back to the earliest inventors in the rubber industry, including British researcher Thomas Hancock in the first half of the nineteenth century. Hancock, often referred to as the "father of the rubber industry," was the first to develop a variety of new machines to fabricate rubber goods. His first invention was a masticator, in 1820, which made it possible to create new rubber compunds. However, instead of calling his machine a masticator and publicizing his success, he called it instead a "pickle" in order to keep his discovery a secret for as long as possible. This misleading term did manage to throw off the competition, and Hancock was able to keep his secret machine safe until 1832 when one of his workers betrayed his trust.[17]

Secrecy continued to be a way of life in the rubber industry. It became especially prevalent in Akron's only two industrial rubber laboratories, located at the B.F. Goodrich Rubber Company and Diamond Rubber Company. Early industrial research laboratories were on the cusp of a new scientific era in American business. By 1900, industrial research had become an important mechanism for advancing scientific progress and corporate profits.[18] While industrial research in rubber chemistry yielded important scientific advances, industry leaders held them as proprietary discoveries, significantly slowing the overall development of the field. Like all rubber companies, Goodrich closely guarded its research secrets.

Charles Goodrich, the son of founder Benjamin Franklin Goodrich, established the company's first laboratory in 1895 to explore practical problems in rubber compounding. Three years later, he expanded it to become one of the first major rubber laboratories in the world. This was only the second rubber research laboratory in the United States. (Boston Woven Hose and Rubber was the first in 1880.) David Goodrich, the founder's other son, recalled that his father "actually had a series of secret recipes for rubber compounds kept in cypher or code." Benjamin Goodrich was so careful about these compounds that he did not divulge them to the new head of research, who replaced Charles, until he married the sister of the company's president. By the early 1900s, many rubber manufacturers tried to learn this secret information by hiring workers from rival firms, thus stealing the proprietary knowledge and making it "common" property.[19]

Secrets also spread when employees left one firm and started their own companies. In 1894, three brothers who had previously worked in the Goodrich bicycle tire department formed the Diamond Rubber Company. They began operations across the street from Goodrich; in their laboratories they researched ways to separate rubber from scrap goods and to improve the vulcanization process. One example of industrial secrecy concerned vulcanization, which was vital in enabling rubber to become a useful substance. Any improvement to the process was of great scientific and commercial importance. In 1905, Diamond hired George Oenslager, a Harvard chemist, to figure out how to "accelerate" the vulcanization process.[20] Acceleration would not only allow for more rubber production and plant efficiency, but a better process might also help to develop stronger rubber. In 1906, Oenslager devised a way to use organic materials (such as anilines) both to speed up and improve vulcanization. These anilines were toxic, but Oenslager eventually developed a non-toxic thiocarbanilide; the result was a huge commercial advantage for Diamond. To maximize the profits from this advantage, Oenslager kept the process secret and even held back filing patents so competitors would not know how much progress had been made. According to Howard Cramer, a Goodyear chemist at the time, this discovery was successfully kept secret until around 1912 when Diamond merged with Goodrich, nearly six years after its discovery.[21]

Secrecy also affected the early years of what was to become the Rubber Division of the American Chemical Society. This organization was founded in 1909 when Charles Goodrich became chairman of its

India Rubber Chemistry Section. Although the goal of its early meetings was to share rubber research throughout the industry, few of the participants were actually willing to give up any of their own secrets. Goodrich himself was intensely secretive, and the other members also would not risk sharing their intellectual property in an open forum. These early rubber section meetings were defined more by reserved silence than open communication. A 1911 report in *India Rubber World* stated bluntly that in the first two meetings of the India Rubber Section there was "not yet sufficient evidence of real cooperative effort among rubber chemists." Many believed that because of the prevailing secrecy, the "methods of analysis of India rubber are in almost a chaotic condition." The journal article concluded that the only way for the field to advance was for the chemists to "get together without too many padlocks on their lips."[22]

If the nascent rubber industry was to continue to grow, a university or college would need to develop specialized coursework for chemists and engineers who wanted to study the science of rubber. Cramer, a Goodyear chemist, believed the rubber industry's penchant for secrecy was detrimental to the industry as a whole. He wrote, "This tendency created the need of some central institution free from all industrial connection to provide special training for chemists in the science of rubber technology."[23] A university could pull back the veil of secrecy, an impossibility in the industrial world. However, despite the growing need, few university professors in the world lectured or provided laboratory experience in this emerging field.[24] This situation was soon to change. Only a few miles from the rubber industry's center was a college with a chemistry professor who had the revolutionary idea to actually start teaching about rubber. Within a decade, this idea would play a major role in opening up the industry, so much so that at the first meeting of the Rubber Division of the American Chemical Society in 1919 everyone believed that the "old bogey of secrecy was at an end."[25]

CHEMISTRY ON THE HILL

In 1870, the same year that B.F. Goodrich arrived in Akron, the Universalist Church, a liberal denomination, set out to establish a new

Buchtel College Campus (1880). *Akron Argo* (1880). *University of Akron Archives.*

institution of higher learning in Ohio.[26] After exploring possible sites in Kent, Oxford, and Mt. Gilead, the church's leaders finally chose Akron, in part due to the persuasion of philanthropist John R. Buchtel. A member of the Universalist Church, Buchtel had earned his fortune by making Akron an important center in the farm machinery industry. He used his wealth to provide a significant portion of the endowment for a new liberal arts college in Akron.[27] After scouting out several sites for the school, Avery Spicer donated the soon-to-be abandoned Spicer Hill Cemetery to the Universalists. This cemetery was situated on high ground, looking west toward the city of Akron, which lay in the nearby valley along the Ohio Canal.[28] In 1914, Akron's "hills and valleys" were called "one of the most picturesque places in the middle West."[29] The school took the name Buchtel College to honor not only John R. Buchtel's funding, but also his personal attention to day-to-day operations. By the end of his life, Buchtel had donated almost all of his wealth to the school, a total of nearly half a million dollars.[30] When classes first opened in 1872, there was one building on the campus, seven faculty members, and forty-six students.[31] One local minister said that they had built a " 'Castle in the air' and called it a Universalist Academy."[32]

At Buchtel College, students were able to select one of three broad areas of study: the Classical Course, the Philosophical Course, and, beginning in 1874, the Scientific Course.[33] Bachelor of Science degrees were awarded to students whose coursework included "wide training in Mathematics and the Sciences" as well as numerous other electives. Students learned Latin or German, mathematics (arithmetic, algebra, and geometry), English (composition, grammar, rhetoric, and literature), history (United States and general history), political science, and drawing. The science courses themselves consisted of natural science (physiology, physics, and physical geology) and chemistry.

In the nineteenth century, college chemistry courses were a new phenomenon in the United States. Those fortunate enough to attend a college or university typically acquired a very broad intellectual background, and few had the option of pursuing a specific degree, such as biology, physics, or chemistry. Narrow academic specialization was not considered an important or worthy intellectual pursuit. Ralph Waldo Emerson symbolized the prevailing ideal of the general scholar. In 1837, he wrote, "Man is not a farmer, or a professor, or an engineer, but he is all." Emerson said that in a specialized society, a mechanic was reduced to a machine, an attorney became a statute book, and a sailor was transformed into a simple rope on a ship. As a result, the sense of the whole was lost. Emerson thought that these specialized men were like "walking monsters—a good finger, a neck, a stomach, an elbow, but never a man."[34] Thus, until the mid-nineteenth century, there was a strong reaction against specialization in the learned professions. After the Civil War, this situation changed dramatically. Herbert Spencer led the charge for specialization as natural and universal progress. He believed that just like organisms that evolved into more and more complex parts (such as the eye), so, too, should human intellectual activity evolve.[35] Spencer saw the "mutual dependence" that specialization required as the fundamental characteristic of all successful organizations (biological, social, or political).

The result of this changing educational philosophy was an increase in specialized areas of study. Johns Hopkins became the first American university to offer a course in chemical research when it was founded in 1876.[36] The school was modeled on the principles of the great German

Professors at Buchtel College (1889). *The Buchtel* (1889). *University of Akron Archives.*

universities of the day, which espoused an academic blending of teaching and hands-on research, as well as a close relationship between chemistry and industry.[37] This idea spread to other leading American universities, and soon a new crop of young professors were anxious to incorporate these methods in their own classrooms. One of these young professors was Charles M. Knight. In 1875, Buchtel College hired Knight to teach its broadly conceived natural science courses—but Knight had his sights set on becoming a specialized chemistry professor and implementing a laboratory learning environment like the one being developed at Johns Hopkins. It would take Knight more than twenty-five years to achieve his goal.

Charles Knight was born in Dummerston, Vermont on February 1, 1848, and spent much of his youth working on his father's farm.[38] When he was sixteen, he left home to attend preparatory school at Westbrook Seminary in Deering, Maine. For the next few years, he returned home to work in the fields each spring because, due to the Civil War, there was a severe shortage of labor. But his thoughts were never far from his studies. While taking work breaks during the planting season, he would sit on the plow-beam and study his Latin.[39] In 1868, after completing preparatory school, he enrolled at Tufts College. To support himself during his schooling, he traveled west for two years to serve as a civil engineer, surveying for the Union Pacific Railroad.[40]

Back at Tufts, Knight finished his degree in engineering in 1873 and then became an assistant principal at Goddard Seminary in Barre, Vermont.[41] One year later, he attended Harvard and MIT for two semesters of postgraduate work in science. But his passion was higher education and service both to a college and its students. In September 1875, he arrived on the relatively new Buchtel College campus as a professor of natural science and quickly began making plans to improve the scientific curriculum. Knight knew that lecturing, reading, and reciting could not provide sufficient scientific training for his students. He believed they needed practical experience and was a strong believer in the laboratory as a place of learning. As a result, in 1876, he personally constructed the campus's first laboratory in the basement of Buchtel Hall, the same year Johns Hopkins introduced the first chemical research course in the nation. He didn't limit his scientific investigations

Charles Knight in 1875, the year he arrived at Buchtel College. *University of Akron Archives.*

to Buchtel College. The *Summit County Beacon* reported in 1883 that "Knight is making some very fine improvements to his home"—three rooms that he turned into a laboratory.[42]

Buchtel Hall was the centerpiece of the new Akron campus and was, in 1879, described by the *Washington Post* as "one of the finest college buildings in the West."[43] The building was a massive five-story structure perched high on a hill overlooking the city. Despite the impressive educational structure, Knight recalled that when he arrived, "There were no arrangements for students to make tests and try experiments."[44] So, during the 1876 Easter recess, he took a discarded table from the kitchen, installed some new washbowls, and connected drain pipes to them. This enabled him to teach a course in quantitative chemical analysis to ten students at a time. Knight then outfitted a second table with a variety of lamps and blowpipes so that students could examine ores and minerals. Knight discovered that students "took such interest in this blending of hand and head work" that he soon moved his tables into a larger room in the basement to accommodate more pupils. The college catalog announced that the new laboratory was "open to students, [and] well furnished with apparatus for making chemical experiments and analyses." During this time, Knight also worked to change the prevailing opinion among students that scientific courses were not as rigorous as the classical courses. Many believed they "did not take high rank for cultural value and mental discipline as compared with the classics."[45] By the 1890s in the United States, chemistry was still considered a "non-professional occupation," and there were very few practicing chemists in America.[46]

Knight's first step in changing this perception was to further his own education. In the early 1880s, he briefly returned to Tufts to earn his master's degree. When he came back to Buchtel College in 1881, he became a professor of physics and chemistry.[47] "We think the college fortunate in securing the services of so fine a scholar," wrote one local reporter.[48] His second step was to continue his struggle to improve the college's laboratory facilities. Attempts to construct a new scientific laboratory began in 1892. The college received $30,000 in pledges from three donors and construction started on a science building one year later. However, due to a severe economic panic, the project was halted, and the bare foundation was left unfinished. The college endured further problems in 1896 when Buchtel's president, Reverend Orello Cone, resigned, and the *New York Times* reported that Cone questioned whether the school would ever be financially prosperous. Buchtel College's own Board of Trustees debated whether the school

Knight's early chemical laboratory and lecture room at Buchtel College (1894). *The Buchtel* (1894). *University of Akron Archives.*

could continue to function.[49] Charles Knight fought hard to ensure that it would. After Cone's resignation, in 1896, Knight took over as president of the college, becoming the only layman to head the school in the nineteenth century. He soon discovered that he much preferred chemistry to administration and returned to his laboratories after one year. But he helped ensure the survival of the school during this difficult period.

The school suffered another devastating setback a few years later when a tragic fire destroyed the centerpiece of the school.[50] On December 20, 1899, a fire of unknown origins broke out in the attic of Buchtel Hall, and the massive building burned to the ground in just three hours. Flames lit up the night sky, and since Buchtel Hall was built on one of the highest points in the county, the fire could be easily seen from the neighboring cities. While students were able to save some of the books from the library and some personal belongings, only one part of the chemistry department survived—a small fifty-dollar platinum dish. The chemistry laboratories were relocated to Crouse Gymnasium.[51] This multipurpose building boasted an indoor running track, exercise equipment, and the school of music, complete with a pipe organ. It also doubled as the school's chapel. All chemistry work was

The "Ruins of Old Buchtel College." This photograph was taken in 1899 and shows the building's skeletal remains following the tragic fire. *The Buchtelite* (May 1909). *University of Akron Archives.*

conducted in the basement in six specially equipped rooms. The inorganic chemistry laboratory took over the space formerly used for the women's locker room and showers. The laboratories had multiple desks for students, each with drainage, gas, and hot and cold water for performing experiments. The entire laboratory was well ventilated, with wall flues removing most of the noxious fumes from the area.[52]

Eventually a new Buchtel Hall was built. The new building, though smaller than the original, had a beautiful classic design with marble steps leading up into the front doors. Once inside, a grand, open, winding double-staircase beneath a skylight welcomed visitors. This new building housed the physics laboratories, with six rooms on the first floor devoted to training students in experimental physics and natural sciences. By 1900, seven buildings made up the entire school on its six-acre campus.

At the start of the twentieth century, Knight's chemistry courses continued to gain popularity despite the limited laboratory facilities. Completing the chemistry curriculum required one year.[53] The first

Crouse Gymnasium (1894). The chemical laboratories were moved to the basement of this facility after the Buchtel Hall fire in 1899. *The Buchtel* (1894). *University of Akron Archives.*

Campus view (1904). The new Buchtel Hall, rebuilt on the same location, with Crouse Gymnasium in the background. *Buchtel College Catalogue 1904–1905. University of Akron Archives.*

semester included an eight-hour-per-week course in general chemistry, where students studied inorganic chemistry, performed and demonstrated laboratory experiments, and manufactured one or more salts. The second eight-hour chemistry course was quantitative analysis, where students studied simple salts, minerals, coal, limestone, milk, water, and foods. This course included another eight-hour chemistry laboratory to apply the knowledge learned from books and lectures. The second semester included a five-hour-per-week class in organic chemistry, an eight-hour-per-week course on inorganic preparations such as common metals and acids, and a ten-hour-per-week quantitative analysis class. The quantitative class enabled students to select an area of industrial chemistry that interested them and an opportunity to visit local manufacturing operations in Akron to gain hands-on experience.

Notably absent from the chemistry courses was the topic of rubber. By 1900, Akron was well on its way to establishing itself as the rubber capital of the world. And yet there was no place other than the shop floor or the industrial research laboratory to further knowledge in the field. Knight welcomed and fostered a close interaction between the rubber industry, his students, and his laboratory. He began to work to bring these two worlds together by introducing the problems of the rubber industry into the classroom and academic research into industry. He saw his college as a vital link in the basic research that would help ensure the continued strength and growth of the American rubber industry.

Knight himself had long been an important resource for the Akron rubber industry, serving as a consultant to companies like Goodrich to help solve problems relating to the chemical composition, structure, and properties of rubber.[54] He also advised the scientists at Diamond Rubber on specific problems they were having in their laboratories.[55] Knight consulted with the other rubber companies as well, attending lectures at Firestone and testing rubber samples for Goodyear.[56] Knight's graduates, who sometimes took leading positions in rubber laboratories, also increased his influence. For example, in the 1890s two of his students, Arthur Warner and John Thomas, joined the fledgling Goodrich rubber laboratories.[57] Starting in 1902, students could elect to visit rubber factories as part of the quantitative chemistry course. Hezzleton Simmons later wrote of his mentor that he had a "keen appreciation of the industrial demands of his city."[58] The *Christian Science Monitor* published an article discussing Knight's legacy, stating that his "Chemistry department serves a city."[59]

By 1902, Knight stopped teaching physics courses to devote all of his time to chemistry. That year he began developing plans to improve the teaching of rubber chemistry. But Knight knew that if he wanted to teach this topic most effectively, his students would need better practical instruction. His laboratories were not sufficient to provide this experience, and a significant new facility was required. To the rescue came one of the nation's wealthiest philanthropists—Andrew Carnegie.

ANDREW CARNEGIE'S GIFT

In late nineteenth-century America, several industrial titans emerged who were as devoted to giving away their massive wealth as they were to making it. Andrew Carnegie was one such giant. He founded Carnegie Steel (later known as U.S. Steel) in 1892 and transformed it into one of the largest and most powerful corporations in America. After earning his fortune, he became a noted philanthropist, making significant institutional donations to libraries and universities. From 1881 to 1919, he gave $56 million to build over 2,500 libraries throughout the world, and 111 in Ohio alone.[60] The heads of Buchtel College thought that, with Carnegie's support, they would finally be successful in establishing a scientific centerpiece at their school. Carnegie responded to their proposal with interest, and in 1905 his associates initiated a six-week investigation at Buchtel College to determine if it met his high standards for a gift. Officials at the school anxiously awaited his decision. When the local papers mistakenly reported that the funds were on the way, Charles R. Olin (class of 1855), who was secretary of the college, had to announce that the "article in the news-paper concerning the Carnegie gift is premature."[61]

Buchtel College President Augustus Church contacted Carnegie's secretary numerous times, but still no official word came. Finally, several weeks later, the philanthropist confirmed that he would be donating $25,000 for the construction of a new science laboratory. Olin wrote, "Mr. Carnegie's offer has given us renewed enthusiasm. It will be a splendid thing for the College when we get the new building. It will give us the much needed room for our Science Dept. especially chemistry."[62] Students cheered the news when they were told at the first chapel meeting after Easter recess. President Church gave everyone a half-day holiday and personally supplied the kerosene for a celebratory bonfire. An additional $25,000 was required for the building and other associated costs, and the Buchtel College staff worked tirelessly for three years to secure commitments from other private donors.[63]

Ground was broken in 1908, and, exactly one year later, in April 1909, the construction of the new building was completed. In May, the Board of Trustees unanimously selected a name for it, proclaiming: "In recognition of his long and distinguished services, the new building [will] be named the Knight Chemical Laboratory and the name be carried upon a suitable tablet upon the building."[64] This was the first building named on campus to honor a faculty member. Charles Knight's salary was raised from $1,500 to $2,500 per year.

Briggs and Nelson Architects built the $30,439, eighteen-room, three-story brick laboratory. Briggs, an 1899 graduate of Buchtel College, was a noted Cleveland engineer and architect. The building's first story included Knight's private office and laboratory, along with a lecture room, general laboratory, and preparation room. The second floor housed a library, physical chemistry room, qualitative laboratory, and water analysis room. It was in the basement of this building that rubber earned its permanent place at the college. Nestled among rooms devoted to electrolysis, acids, explosives, and gas analysis was a fourteen-by- eighteen-foot "rubber laboratory."[65] The building was unveiled to the public in June 1909, when Buchtel College held a commencement reunion. The centerpiece was the grand opening of the Knight Chemical Laboratory. Knight was the honored guest, and the event ended with "old English dances on the campus."[66]

Once the building was completed, Knight worked to equip it with the latest laboratory devices. Knight and Charles Olin saw to it that local business leaders knew their needs and solicited them for donations. Knight made careful notations of each of these gifts in his daily

Knight Chemical Laboratory (1911). *Tel-Buch* (1911). *University of Akron Archives.*

diary. For example, one entry reads, "The calendar machine donated by Mr. Adamson was sent inside the lab."[67] In another entry, he said, "I went over to the Biggs Boiler Works to solicit a vulcanizer for the rubber laboratory."[68] There were numerous other donations due to Knight's tireless work: Firestone gave a press vulcanizer; the rubber washer and mixing mill came from the Farrel Foundry and Machine; and the steam generator with an automatic regulator was a gift from Williams Foundry and Machine.[69] According to Maurice Morton, who later headed the University of Akron's polymer science program for a quarter century, the importance of Knight's successful efforts in securing apparatus for the laboratory could not be overstated. Morton wrote, "This was a critical problem . . . since such machinery was (and still is) essential in the study of rubber compounding and vulcanization and represents an expense which Buchtel College could ill afford."[70] With the new building completed, the laboratory apparatus mostly in place, and his ties to the local rubber industry increasing, Knight was ready to begin teaching his new course in rubber chemistry.

THE WORLD'S FIRST RUBBER CHEMISTRY COURSE

When Knight first offered his course, secrecy still dominated the rubber industry. According to William C. Geer, vice president of research at Goodrich, "Mystery has surrounded rubber factories [and] secretiveness has been the watchword.[71] In 1910, Geer had met with Knight and discussed the role the rubber technology course would play in relaxing some of the secretiveness. Both were hopeful that Knight's educational experiment would have a significant impact. That same year, Knight invited George Oenslager, the chemist at the Diamond Rubber Company who had discovered the process of using organic accelerators in the vulcanization process, to meet him at the new Knight Chemical Laboratory. There the two talked about "circular letters to the rubber workers" announcing the learning opportunities at Buchtel College.[72] Oenslager had been a champion of secrecy. But after his meeting with Knight, he realized the immense value of an independent, academic center for rubber research.

Early advertisements for the rubber course aroused the immediate interest of many Akron rubber workers. In 1909, *The Buchtelite* (the Buchtel College newspaper) reported, "The rubber course is attracting special attention. Applications have been received from several who wish to take up post graduate work and many workmen from local factories have expressed their intentions of taking the course."[73] Before the opening of the Knight Chemical Laboratory in 1909, Knight had been introducing more elements of rubber chemistry into his courses, but the first official course in the subject, "The Chemistry of India Rubber," was not offered until September 1909. [74] Knight described it to prospective students as a "course of training in the chemical examination of crude and cured rubbers."[75] This advanced class was available to students only after they had finished courses in both quantitative analysis and organic chemistry. Although Knight made only casual mention of the course in his diary, stating, "Class in Rubber Chem. met for 1st time," these were the first such courses taught by a university professor anywhere in the world.[76] It was the only one of its kind for over twenty years.

Knight was nearing the end of a long career, having spent thirty-five years at Buchtel College by the time he taught that first course in rubber chemistry. But for the final three years of his tenure, he established

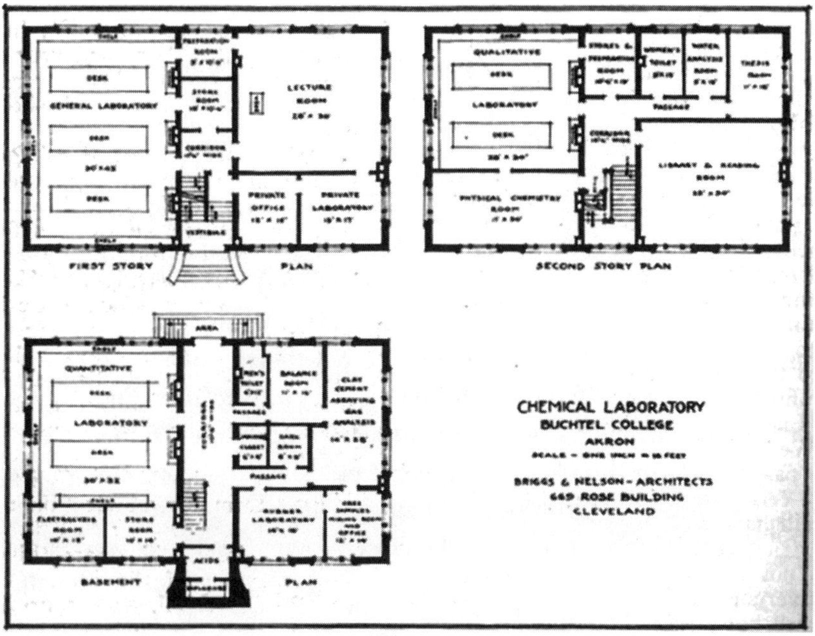

Knight Chemical Laboratory floor plan. *The Buchtelite* (April 1909). *University of Akron Archives.*

his greatest legacy by focusing on rubber specialization. Knight envisioned the teaching of rubber at the college as far more than just a simple course. From the beginning, it was actually conceived as a subspecialty within chemistry to which students could devote up to two full years. The first year was for traditional coursework and laboratory research where students worked eight hours per week. The first semester explored various crude gums, solvents, and methods to analyze "soft-cured" rubbers. The second semester covered theories of vulcanization, methods of compounding and curing rubber samples, and practical application. The second year was designed for students who planned to continue their research in rubber chemistry, either as field research chemists or as academician. Knight always emphasized the importance of hands-on work. As a result, his students processed all of their rubber themselves. This included using mixing rolls, calendars, and vulcanizers. Knight also enabled students to tailor the courses to fit their needs. While two years of instruction was available, students could register for as much or as little of this as they wanted, and certificates describing the work each student completed were available whenever they decided to leave the program.[77] Knight also made certain that the library contained adequate resources for his students. In 1910, nearly one-third of the new books acquired at the college library that year focused on the subject of rubber.[78]

But trouble was again on the horizon for the college. Its well-regarded president, Augustus Church, had died in November 1912 of pneumonia.[79] It fell to his successor, Parke Kolbe, to deal with increasing financial pressures. Kolbe's response was to relinquish the college's assets to the city of Akron. He knew that if the city did not take over control of Buchtel College, its doors would be forced to close. Kolbe had to convince city leaders that the school was more than just a center for those of the Universalist faith, but that it had become an important academic institution and was in Akron's best interests to save. By August, Akron's mayor and the city council passed an agreement to transfer ownership of Buchtel College and all of its assets to the city, and the school became the Municipal University of Akron.

Charles Knight had been instrumental in the transition and an important proponent of the change. But that same year he retired from the University of Akron after thirty-eight years of service. The new board of directors stated: "It is recommended that in view of his long and faithful services to this college, Dr. C.M. Knight be retired with the title of Professor Emeritus of Chemistry and Dean Emeritus of the Faculty with an allowance of $750 per year."[80] It fell to his successor, and former student, Hezzleton Simmons, to take the three-year-old course in 1913 and establish a national reputation for it.

SIMMONS AND RUBBER CHEMISTRY

On July 30, 1906, Hezzleton Simmons received an important letter at his home from Augustus Church, the president of Buchtel College. In the letter, Church explained that Charles Knight was anxious to find a student to serve as his new research assistant in the chemistry department. Knight had always been impressed with Simmons's work ethic and believed he was the best person to fill the job. Church offered Simmons, who was starting his junior year, the research position, tuition

Hezzleton Simmons (back row, second from left) as a student at Buchtel College. Here he is standing with the "Curtis College girls" in 1908. *University of Akron Archives.*

remission, a stipend of $3.50 per week, and free chemical laboratory fees up to $14.00. Simmons quickly accepted the offer and one month later began working closely with his new mentor, Charles Knight.[81]

Simmons was born in 1885 in Lafayette Village, Ohio. Soon after, his farming family moved to Leroy, Ohio, a city that had been settled by Universalists. In 1903, he graduated from Leroy High School and decided with his parents that higher education would provide him a better life than the farm would. So his father hitched his best horse to his wagon and took his son and his belongings to a small room on Union Street, near Buchtel College.[82] Simmons later dedicated his influential book, *Rubber Manufacture*, to his parents who, he said, "by their sacrifices made it possible for me to obtain an education."[83] Simmons immediately selected the Scientific Course and excelled in chemistry. He was twenty-one years old when he received the letter from Buchtel College's president.

Simmons served as Knight's research assistant at a time when the college was growing. Charles Knight invited Simmons to overturn the

Hezzleton Simmons's 1908 graduation picture. *The Buchtel* (1908). *University of Akron Archives.*

first two shovelfuls of dirt with him at the 1908 groundbreaking of what was to become the Knight Chemical Laboratory. Simmons recalled, "They were very difficult to remove because they were taken from the center of the only tennis court that the college then owned."[84] That same year, Simmons graduated with his Bachelor of Science degree from Buchtel College, and Knight helped to get him accepted at the University of Pennsylvania, where he continued his studies in chemistry. He was fascinated by rubber chemistry and spent two summers investigating production processes at the Goodyear factories.[85] In 1910, he was called back to Buchtel College to become an instructor in chemistry.

Knight concentrated on his duties as head of the chemistry department and gave most of the lecturing to Simmons. The only exception was the new course in rubber chemistry that Knight pioneered and wanted to teach personally.

When Knight retired in 1913, Simmons began preparing for an exciting new era. "Industry," wrote Simmons, "was recognizing that practically all problems were based upon chemical laws and therefore chemists in greater numbers must be trained to do its work."[86] The most dramatic area of growth was in the field of rubber chemistry. With Knight's legacy of establishing a dedicated laboratory and teaching the first course of rubber chemistry completed, it was up to Simmons to strengthen and expand this program, as well as continue to forge new ties with industry. Knight had only taught the rubber course for three years, and now Simmons had the opportunity to make it his own. He began by working to bring wider attention to rubber chemistry at the University of Akron. In an era in which secrecy still surrounded knowledge in the rubber industry, Simmons opted for a unique strategy—he decided to serialize and publish all of his lectures in *The Rubber Age*, a leading rubber journal. Simmons's decision to open the doors of a rubber laboratory in the journal's pages just months after America had declared war on Germany helped establish a new era of academic freedom in rubber research.

Beginning in July 1917, *The Rubber Age* published all of Simmons's lectures. The journal's editor, Herbert M. Williams, believed that this would "form the most complete up-to-date and authoritative treatise that has ever been published on the subject."[87] *The Rubber Age* was an excellent place in which to publish these lectures. In 1917, it had the largest paid circulation and newsstand sales of any rubber publication of its day.[88] Williams also believed that Simmons was the best person in the country to publish this material. He said, "No man is better qualified to do this work than he, for his intimate association with the industry, his standing as a practical rubber chemist, and his years of teaching the subject have thoroughly fitted him for this work."[89] The twenty-one articles presented in the journal were nearly identical to the lectures Simmons delivered in his lecture hall at the University of Akron, such as the history of caoutchouc, rubber plantations and their development, theories of vulcanization, and physical testing of compounded samples.[90] Because this was the first time such a complete story of rubber chemistry had been printed, Simmons's articles immediately became popular. His articles would later be compiled into a book, *Rubber Manufacture*, and published in 1921.[91] After the first lecture was published, *The Rubber Age* received over one thousand new subscriptions from readers associated with almost every rubber company in existence.[92]

The publication of Simmons's lectures coincided with the United States' entry into World War I. In complete contrast to the situation the nation would face at the onset of World War II, most observers believed in 1917, the rubber industry would not be affected by war. Henry Firestone confidently proclaimed, "America's entrance into the war will not materially affect the rubber industry."[93] After all, access to rubber plantations in the Far East and the supply of natural rubber were not threatened by the conflict.

Nevertheless, war did alter the direction of the chemistry department's work at the University of Akron. World War I has sometimes been called the "chemist's war" because these professionals dominated the scientific experts.[94] Army and navy projects utilized almost forty different college and university laboratories throughout the United States, and Simmons volunteered the University of Akron's rubber expertise. In the spring and summer of 1917, the university created a special rubber school in conjunction with the Committee on Education and Special Training of the War Department. The goal of the school was to train soldiers in the art of tire repair and vulcanization, skills that would yield important benefits on the battlefield. Over five hundred soldiers trained at the University of Akron. They were quartered in Crouse Gymnasium and ate in a temporary mess hall constructed on campus specifically for them.[95]

In the fall of 1918, the Department of Chemistry was also involved in establishing an intensive course for the Chemical Warfare Service. The U.S. Bureau of Standards established a branch at the Knight Chemical Laboratory specifically for the purpose of testing rubber tires purchased by the government. Approximately 70 percent of all tires purchased for the war by the government were tested at the University of

OUR RUBBER LAB

The Charles Knight chemical laboratory. There is a two-pan balance in the center of the picture. (*top*) To the left of it is an autoclave (which was able to heat water to temperatures higher than 100 degrees Celsius). At the far left hand side of the image is a compression tester. The glass structure toward the right is a fume hood. Pictured here (*bottom*) are two separate lab mills driven by one motor located on the back of the table to the left. Both pictures were taken when the rubber laboratory opened in 1909. *Tel-Buch* (1918). *University of Akron Archives.*

Akron laboratory. Arnold Smith, a former Buchtel College student, ran the laboratory, and the government used twelve other graduates of the rubber chemistry course as experts.[96] In the summer of 1919, the lab was shut down, and the University of Akron bought all of the testing equipment for use in its rubber chemistry course.[97] Regular instruction in rubber chemistry resumed in January 1919.[98]

After the war, Simmons strengthened the ties between the University of Akron and the rubber industry. Like Knight, he always included field trips to local rubber companies to complement his lecture and laboratory assignments. Simmons said, "It should be noted . . . that great value to our men is obtained from frequent visits to the large factories here."[99] For example, Simmons might lecture on the topic of the washing of rubber. Students would explore this in their own laboratories. Finally, to understand the process in action, Simmons would take them to visit the wash rooms of several large rubber companies. Simmons believed that the "cooperation of the manufactures of our city" was one of the great learning opportunities for his students.

His close industry ties also resulted in the beginnings of the master's degree program in rubber chemistry. In the fall of 1917, Goodyear and Firestone created fellowships for the study of rubber chemistry at the University of Akron.[100] Those eligible were chemistry graduates from other schools who wanted to specialize in rubber chemistry. The fellowships paid $300 per year (this eventually was increased to $500 and then $1,000 per year), and recipients were exempt from all tuition and laboratory fees. In return, they had to serve nine hours each week as laboratory supervisors in organic chemistry. In 1929, B.F. Goodrich began offering a similar fellowship. These became very popular, and in 1931 there were over sixty applicants to fill the three positions.[101] Since these students were already graduates of other chemistry programs, they were eligible for a master's degree once they completed their year's work at the University of Akron.[102] The students then began work at whatever company's fellowship they had held. By 1925, the name of the course and program was changed to "Chemistry of Rubber Technology" and remained the same for the next forty years.

Over time, the program offered more complex analytical techniques and developed more rigorous hands-on operations. Simmons believed his graduates were well-prepared to enter industry, as they had all spent time working on practical problems in his rubber clinics. Rubber companies throughout the world submitted their problems, and Simmons assigned students to work on solutions. One standard assignment for all students was dissection. While medical students dissected animals and cadavers to understand better how an organism was put together, Simmons had each of his students dissect a tire. The students then analyzed the rubber, computed its formula, and were responsible for developing a new compound as good as or better than the one in the original tire. This required the student to go into the mill, weigh and mix the ingredients, compound the rubber, and then test the physical properties. Simmons would suggest ways to improve the compound by adding various chemicals to it. The final step was to compute the actual volume cost of the rubber. The process was designed to take the student through the entire process of manufacturing commercial rubber.[103]

Hezzleton Simmons (1924). *Tel-Buch* (1924). *University of Akron Archives.*

Simmons did not neglect the educational needs of those already working in industry. It would have been impossible for these men to leave their factories during the day to take courses.[104] As a result, in 1928 Simmons helped to establish the evening college program at the University of Akron. While more broadly conceived than simply as an opportunity for rubber workers to acquire advanced knowledge, the rubber course played an important part in these evening offerings. In the early 1930s, it was presented as a lecture-only course with no laboratory work.

The reputation of the rubber chemistry program at the University of Akron soon spread internationally. In 1929, three men from Asia arrived at the university. Hideo Kaneko, a graduate of Yokohama University, and Yasutoshi Uchiyama, a graduate of the University of Tokyo, came from Japan to study rubber chemistry with Hezzleton Simmons. Likewise, Ching F. Kuo, who was born in the Shansi Province in China, received his bachelor's and master's degrees from the University of Illinois but came to Akron because it was still the only university offering advanced coursework in rubber chemistry.[105] The reputation of the rubber program spread to Europe as well. In 1931, William Hohl of Zurich and George Stepanek of Prague came to the University to take the special course in rubber chemistry offered by Simmons.[106] In 1934, in appreciation of the training in rubber chemistry received by Japanese students over the years, Professor Takeo Kimishima of the Kyusha Imperial University returned to Akron, where he had graduated in 1919, to be reunited with his mentor, Hezzleton Simmons. Kimishima presented Simmons a gift of three Japanese swords, heirlooms dating back to the medieval empire.[107]

Hezzleton Simmons had done a great deal to advance the study and teaching of rubber at the University of Akron. In 1933, he moved out of his beloved chemistry department to become the president of the university, a position he held until 1951. Walter Cook (who had come to the university in 1926 as an assistant professor) became the head of the chemistry department.[108] Howard Cramer of Goodyear began teaching the rubber chemistry course in 1929. Cramer had been one of Simmons's students and later earned his Ph.D. at the University of Wisconsin.[109] "I am an inheritor," said Cramer, adding that he hoped to continue Simmons' tradition of excellence.[110]

Perhaps most prominent among the many Akron students who went on to careers in the rubber industry were two graduates who guided Firestone: J.W. Thomas and Lee R. Jackson. Thomas graduated from Buchtel College in 1904 and founded the first Firestone laboratory in 1908. He discovered a formula that improved rubber compound qualities. He went on to serve as Firestone's president from 1932 to 1942 and was the chairman of its board until 1946. He was also elected a board member at the University of Akron in 1919, serving as chairman from 1926 until 1942.[111] Jackson also had a distinguished career at Firestone. He graduated from the university in 1913, although one year earlier he had taken a position as a salesman for Firestone. He slowly moved up the ranks of the company, becoming vice president in 1929, executive vice president in 1941, and president in 1948.[112] He retired as vice-chairman of Firestone in 1961.[113] Like Thomas, he exemplified the academic/industrial link between the University of Akron and the surrounding rubber industry. He gave back to the university by serving on its Board of Directors for many years and was its chairman from 1951 to 1954.

One mark of a successful teaching career is the quality of the students produced. Knight and Simmons both contributed to the training of a distinguished list of alumni in rubber chemistry. Over his nearly forty-year tenure, Knight inspired a number of his students to achieve important contributions in their fields. For example, Joseph H. James (class of 1894) became the head of chemistry at the Carnegie Institute of Technology; Ralph E. Myers (class of 1901) became the chemical director for Westinghouse Lamp Works in Bloomfield, New Jersey; and Hezzleton E. Simmons (class of 1908) succeeded Knight after his retirement and eventually became president of the University of Akron. By 1939, the university had graduated 150 students from the course, and 115 actively worked in the rubber industry.[114]

CONCLUSION—THE KNIGHT AND SIMMONS RUBBER LEGACY

Despite the significance of the rubber chemistry course at the University of Akron, no other educational institution followed its lead for over two decades. It was not until 1934 that the University of Southern California at Los Angeles began offering an evening course of twelve lectures in rubber chemistry under R.B. Stringfield.[115] In the mid-1930s, Millard Fillmore College at the University of Buffalo began offering a similar evening course.[116] Still, the University of Akron remained unique. As late as 1939, the university promoted its rubber chemistry course as the only one in the world that combined lectures and laboratory work on procedures actually used in rubber factories.[117] By opening the doors of academia to rubber chemistry, Charles Knight and Hezzleton Simmons had established a pioneering academic program.

Throughout his retirement, Knight spent his summers in Akron and winters in Coral Gables, Florida, pursuing his passions in photography, reading, and gardening. He remained an advisor and supporter of the university and an honorary member of the American Association for the Advancement of Science.[118] In 1937, he donated a plat of land (83 by 62 feet) to the university, extending from Buchtel Avenue to Carroll Street. The land was used for expansions to the Knight Chemical Laboratory and the library.[119] Knight died in 1941, at the age of ninety-three, at his Coral Gables home.[120] In 1920, reflecting upon Knight's career, his student Joseph James, head of chemistry at the Carnegie Institute of Technology, wrote, "He was one of the first rubber chemists in the country, having done chemical work for the Goodrich Company years before the management of many rubber companies realized that a chemist is a necessity rather than an expensive luxury. . . . I can testify that Buchtel students were indeed fortunate to have such a teacher in this science."[121]

Simmons also left a strong legacy in rubber chemistry. He had helped to popularize Knight's course; he had strengthened the university's industrial ties and established the master's degree program; he and his rubber chemists had worked with the government during World War I; and he had expanded the department's reputation and brought international attention and participation in the program. Through the efforts of Simmons, and his mentor Charles Knight, the University of Akron had played a seminal role in the birth of rubber chemistry and in defining a new discipline. But, by 1941, the United States was on the brink of World War II, and the nation's leading rubber chemists turned their attention to a rubber crisis that threatened the outcome of the war. If the nation did not establish a synthetic rubber program, and develop

CHARLES M. KNIGHT, A. M., ScD.

Dr. C. M. Knight was born in Dummerston, Vermont a little over one hundred years ago and although he did not reach Buchtel College until he was twenty-seven years old, his entire life and history is surrounded in the legend and fact of "Old Buchtel" and the present University of Akron. He was as it is told in Fifty Years of Buchtel . . . "a giant among giants in teaching at Buchtel". He established and taught and built the Chemistry department here and although he was President for only one year, he could have had that position indefinitely had he so desired. His heart and his love, however, were to be found in the classroom and among the apparatus of his laboratory. In his day, the students and the professors studied for cultural growth as well as practical professional experience, yet he was prominent among the infant industry, rubber, in Akron and helped it to develop. It is not difficult to see how much influence he had through the years and even today for when it was decided to put up a new chemical laboratory, there was no doubt but that it would bear the name of Knight.

The Artist — Miss Lulu Knight
Miss Knight is a granddaughter of Dr. C. M. Knight

Portrait of Charles Knight, drawn by his granddaughter Lulu Knight, c. 1940. *University of Akron Archives.*

A statue of Charles Knight standing on the west side of the Goodyear Polymer Center today. It was dedicated by his granddaughter, Helen Iredell Gulick (class of 1940), in October 1996. The statue overlooks the current Knight Chemical Laboratory. *Photo by Mark D. Bowles.*

one quickly, the future of all the allied nations was at risk. The administrators at the University of Akron and its rubber chemistry program played an essential role in this quest. In confronting this challenge and national emergency, Simmons would leave his strongest and most lasting legacy.

2

THE SYNTHETIC QUEST

1941–1956

"ATTENTION, CITIZENS OF THE UNITED STATES!" WROTE AN MIT professor in 1942. "Our present rubber situation is serious, deadly serious. . . ."[1] With nearly the entire world's supply of natural rubber under the control of Japan, an Allied military that literally moved upon rubber threatened to grind to a halt if a dramatic action were not immediately taken. In response, a consortium of organizations involved in rubber research and production "united in a unique spirit of technical cooperation."[2] A vast network of academic and industrial researchers throughout the United States cooperated in working toward developing a synthetic rubber to replace the lost natural supply. The result was the commercialization of a general purpose synthetic rubber known as GR-S (Government Rubber-Styrene).[3] The program's importance paralleled the Manhattan Project in scope but exceeded it in terms of practical significance for the war.

Of all the government, academic, and industrial organizations that worked toward achieving this rubber milestone, only one location was chosen in 1998 by the American Chemical Society as the site for its National Historic Chemical Landmark. Today, a black and white granite memorial rock sits at the entrance of the University of Akron's Goodyear Polymer Center, recognizing the significance of the United States Synthetic Rubber Program from 1939 to 1945 and its central importance in the Allied victory in World War II.[4] "Had it not been overshadowed by the production of the atomic bomb," wrote several U.S. senators in 1948, "this achievement would undoubtedly have been the outstanding accomplishment of the war effort."[5]

THE RUBBER CRISIS

On April 9, 1942, a fight broke out in the exclusive Alfalfa Club at Washington, D.C.'s Hotel Willard. According to the *Los Angeles Times*, the "sideshow of fisticuffs" took place between two distinguished men in their late sixties, each wearing a suit and tie, out for an evening of dinner and drinks.[6] The disagreement was over a somewhat unusual topic—synthetic rubber. Jesse Jones, the U.S. Secretary of Commerce, walked into the club and began verbally berating Eugene Meyer, the owner of the *Washington Post*. The *Post* had an article alleging that Jones and the Roosevelt administration had failed to develop a synthetic rubber program to meet the growing wartime emergency. The heated exchange escalated when Jones grabbed and began to shake Meyer, knocking his glasses from his face. Meyer responded by taking a wild swing at Jones's jaw. Bystanders, among whom were Supreme Court justices and U.S. senators, eventually separated them.[7] What could have made a seemingly benign topic such as rubber arouse such emotional anger among the Washington elite?

During World War II, rubber had become a material as vital as any weapon and one that could determine the outcome of the war itself. Just a few months before the Japanese attack on Pearl Harbor, the *Wall*

Street Journal discussed the importance of rubber for war, proclaiming, "The army moves on rubber."[8] Tanks required rubber treads. Trucks, airplanes, automobiles, and even gun carriages depended upon rubber for their mobility. The military used rubber to keep airplane gasoline tanks, oil tanks, and fuel lines from leaking, often despite bullet holes in combat. The army also used it to protect troops from poison gas attacks through rubber-reinforced masks and clothing. The quantities required were staggering. A single battleship used seventy-five tons of rubber, and a B-17 Flying Fortress used a half ton. In World War I, the military required thirty-two pounds of rubber per soldier; by World War II, this had increased to 194 pounds.[9] Simply put, without rubber, modern warfare would come to a grinding halt. Furthermore, if one combatant had access to rubber and the other did not, victory would likely go to the one that possessed this essential material.

As World War II escalated, a critical problem emerged for the United States. It was almost entirely dependent upon other nations for its rubber supply, and these sources were being cut off. In 1939, natural rubber represented the United States largest imported commodity (in dollar values). A vast majority of it came from the Far East, and when the Japanese attacked the Dutch East Indies (what is now Indonesia) and Malaya during the winter of 1941–1942, the United States no longer had access to that supply. The Allies found their access blocked to over 90 percent of the world's rubber-producing areas. Only limited amounts of natural rubber were available from elsewhere—41,000 tons from Latin America, for instance, and another 18,000 tons from Africa. This paled in comparison to the nation's need for 655,000 tons per year in times of peace, and 766,000 tons at the start of the war.[10] The United States immediately began a recycling program and protected its stockpiles, but it was clear that it would soon run out of the precious substance. The country was in a state of emergency.[11]

In response, in August 1942, President Franklin Roosevelt assembled a special three-man committee to investigate the "rubber situation." This group included financier Bernard M. Baruch, Harvard University President James B. Conant, and MIT President Karl T. Compton. They concluded, "Of all critical strategic materials, rubber is the one which presents the greatest threat to the safety of our nation and success of the allied cause."[12] To compound the crisis further, while the United States and most of its Allies had lost the ability to import rubber, their chief enemy, Hitler's German army, could produce a synthetic rubber, called Buna-S. The only synthetic rubber available in the United States at that time was DuPont's Neoprene, and there were still numerous scientific questions to be resolved before chemists could produce it commercially. Roosevelt implemented a large-scale governmental plan with the goal of producing over 800,000 tons of a Buna-S type rubber in a massive academic and industrial mobilization.[13] There was no time to waste.

This result was a "production miracle" responsible for 719,000 tons of GR-S synthetic rubber in 1945 alone. Though people most often cite the Manhattan Project and the building of the atomic bomb as the shining example of government-directed scientific research and development, in reality, without the synthetic rubber program, the outcome of World War II would have been dramatically different. In three short years, a new industry was born, and the United States moved from being the largest importer of natural rubber to the world's largest exporter of synthetic rubber.

Despite its significance in the war effort, the importance of synthetic rubber and the role it played in the conflict is not often appreciated. In 1945, just over a month after the Germans surrendered, journalist R.H. Markham noted that even though everyone knew automobile tires were precious and scarce, few realized the importance of rubber in shaping the future of a free world. Markham wrote that rubber is "a dramatic object, closely connected with imperial struggles, world organization, political battles, the social order, and advanced research."[14] Although the synthetic rubber program was a national effort, Akron remained at the center of this political, military, and scientific initiative. Markham concluded that Akron, "with justification, is called the world rubber capital." It was not only the "geographical center of tire production," but also the research required to further the knowledge of rubber was based there. The development of synthetic rubber required a massive, coordinated effort in basic research.

The University of Akron played important scientific and administrative roles in supporting the United States in its quest for the synthetic rubber research program during World War II and after. Scientifically, the university established one of ten U.S. research laboratories—the Rubber Research Laboratory—designed to study problems directly related to the production of synthetic rubber. G. Stafford Whitby, remembered today as "undoubtedly the most honored rubber chemist of all time," headed this research program, which led to important laboratory advances in such areas as "cold rubber."[15] Even President Harry S. Truman noted the importance of this achievement. "Cold rubber," he wrote, "has been found to be superior to natural rubber in most tire treads."[16]

The university also played a key administrative role in the synthetic rubber program. The administrative component of basic research is often overlooked, but for scientists to be successful, they need an organized and coordinated system in which to operate. The University of Akron's president, Hezzleton Simmons, spent much of his time in Washington, D.C., working with the national rubber director to direct the overall synthetic research policy. Back on campus, the War Production Board gave the university responsibility for managing funds for a synthetic rubber Pilot Plant and Evaluation Laboratory staffed by several hundred people. President Norman P. Auburn, who succeeded Simmons, believed the administration of this basic research helped to win the war. He said that these facilities as "operated by The University of Akron, and known as the Government Laboratories . . . provided [a] storehouse of knowledge vital both to national security and industrial development."[17] At the government's bequest, the university administered more than $8 million, more funds than any other single academic, industrial, or governmental organization in the entire synthetic rubber program. Akron, the surrounding rubber companies, and its university were at the center of the synthetic quest.

SEIBERLING, SIMMONS, AND SYNTHETIC RUBBER

Efforts to develop a synthetic alternative to natural rubber began in the nineteenth century, but it was not until 1929 that researchers at IG Farben's Leverkusen laboratory, near Cologne, developed a practical synthetic known as Buna-S (butadiene and styrene). Although the

Germans patented this process, it was not considered a significant breakthrough at the time because the price of the abundant natural rubber was very low during the early depression years. Buna-S initially also had major physical drawbacks: it cracked once in use and it could not be processed on existing rubber machines.[18] Hezzleton Simmons and other leaders of the Akron rubber companies took note of the German advances and, in 1929, also began making plans to establish a synthetic rubber institute at the University of Akron to perfect the process of creating a natural rubber substitute.

Given the university's well-established reputation in the field, it became the logical place for a U.S. institute to guide basic research into synthetic rubber and also to coordinate the efforts of industry. In 1929, officials from the university and local rubber companies developed a plan to create a "fundamental research department manned by experts and the finest equipment [which] would serve the majority of America's rubber manufacturers."[19] While synthetic rubber was one focus, other research would include the vulcanization properties of rubber. The institute would release all experimental results to members of the Rubber Manufacturers Association for further distribution to American rubber companies.

In May 1929, the Rubber Manufacturers Association president, Frank Seiberling, organized a committee to discuss the formation of the institute.[20] Seiberling had founded the Goodyear Tire and Rubber Company with his brother Charles in 1898, but he had then left in 1921 to form the Seiberling Rubber Company. He used his significant clout to attract the heads of the American rubber industry to discuss the new institute. These included James D. Tew (president of B.F. Goodrich), Ernest Hopkinson (vice president of United States Rubber), E.B. Germaine (president of Dunlop Rubber), and Paul Litchfield (president of Goodyear Tire and Rubber).

While Seiberling's committee considered an organizational framework for the institute, Seiberling also established an advisory group to work out the technical details. The advisors were chief chemists from the major rubber companies and Hezzleton Simmons from the University of Akron. R.P. Dinsmore, chief chemist of Goodyear, was chair. Along with Simmons and Dinsmore, other members included

N.A. Shepard (director of research at Firestone), J.W. Schade (director of research at B.F. Goodrich), and W.A. Gibbon (director of research at United States Rubber). Their primary job was to write a report discussing plans for the institute.

These men shaped and controlled the future of the rubber and tire industry, and they recommended, above all, that it was essential that the institute be under the control of the University of Akron. The geographic proximity of the major rubber companies to the university made it a convenient place to perform research, and the University of Akron had been teaching rubber chemistry since 1909; its professors and laboratories were among the best in the country. The university had an added advantage because many of those who held key positions in the rubber industry had actually been trained at the school in its undergraduate and graduate programs. According to their report, "These men have made numerous contributions to the science of Rubber Technology," and they considered housing the institute at the University of Akron a homecoming.[21]

These advisors also recommended that the institute should focus primarily on basic problems in rubber chemistry, physics, and physical chemistry and exclude any work in applied research or development. The hope was that the institute's work in basic science, particularly in synthetic rubber, would stimulate research by major American rubber companies. They made a compelling argument for the institute's necessity: "It should be realized that much of the fundamentals in connection with rubber is, as yet, entirely unknown." The rubber industry, in general, had not committed necessary funds for researching these basic questions. However, they warned, a "great deal of the known fundamentals had been contributed by foreign investigators."[22] This was an unacceptable risk for the United States, given an uncertain future, especially in light of Germany's progress on synthetic rubber and Buna-S, and its past proclivity for war.

These were expensive goals. In 1929, the group proposed a 1930 budget of roughly $200,000 for the construction of buildings at the University of Akron, the purchase of equipment, and the hiring of chief personnel. The new structure was to be located between the physics and chemistry buildings on campus, and all three would be connected with underground tunnels. Ideally, the same architect who designed the Knight Chemical Laboratory would also design this one, so that the buildings would "all be in architectural harmony." The estimated budgets from 1931 to 1933 were roughly $120,000 per year. This resulted in a four-year total of $560,000. In the booming economic environment of the 1920s stock market, these expenditures seemed feasible for all the participating companies.

Although there was great enthusiasm for the institute as well as support for the industrial/academic relationship, everything came to an abrupt end when the economy suddenly collapsed in October 1929, marking the onset of the Great Depression. The rubber industry and tire manufacturers were severely hurt. As Americans stopped buying automobiles, the sales of tires quickly dropped from 47 to 33 million per year, and the rubber industry as a whole experienced a decline in sales from $323 million in 1929 to $171 million by 1932.[23] The result, noted one Akron observer, was a period of "collapsing demand, vanishing companies and the readjustment of a billion-dollar industry on a fairly permanent basis of half size."[24]

Seiberling's hopes for the survival of the institute began to fade in the early months of the depression. In January 1930, he wrote a letter to George Zook, president of the University of Akron, and said, "I regret to report that the committee having the Institute matter in hand did not have a report ready for our Board and consequently no action was taken at our meeting."[25] Seiberling also noted that it was increasingly difficult to get all of the interested parties together to make any firm resolutions about the initiative. The economy worsened, pessimism spread throughout the nation, and the Depression caused significant economic hardships for rubber companies; no funds were available to finance such an extravagant and forward-thinking program, and the first attempt to create an Institute of Rubber Research failed. An institute with this same name would later be formed by Maurice Morton at the University of Akron in March 1956.

The ideas behind the failed rubber institute anticipated one of the central concerns that the nation faced in World War II. With the United States finding itself cut off from the natural rubber supply, Germany achieving major advances in synthetic rubber, and with warfare increasingly dependent upon rubber for military success, the United States needed to find an alternative to natural rubber. Some progress was made during the Depression. DuPont successfully developed Duprene (later calling it Neoprene) in 1931, General Tire in Akron performed research on Buna-N in 1934, and other Akron rubber companies tried to license the IG Farben Buna-S technology. Waldo Semon and Lorin Sebrell, chief scientists from Goodrich and Goodyear, each visited IG Farben in 1937 and 1938, but they could not learn any of the German trade secrets. Upon their return to the United States, each company renewed efforts to create a synthetic rubber from a process not covered under the IG Farben patent.[26] By the end of the 1930s, the United States still imported 95 percent of its natural rubber from Asian plantations. Had the Institute for Rubber Research actually been established in 1930, the nation would have had over a decade's head start and might not have experienced the rubber crisis, which threatened to lose World War II for the Allies.

Simmons (*right*) and Charles Stillman (*left*), a Goodyear vice president (1939). *University of Akron Archives.*

WHITBY, WORLD WAR II, AND RUBBER RESEARCH

World War II changed the lives of every American and shifted the focus of nearly all of the nation's institutions. The University of Akron likewise implemented changes due to the war. Professors taught new courses in aircraft materials and construction, algebra for aircraft workers, and slide rule calculation. The Department of Home Economics began teaching how to cook without sugar and with meat substitutes, while the Art Department began a special program in how to draw and paint camouflage.[27] The most dramatic changes came in the sciences, and most of the university's research facilities were used to support the war effort. While other academic institutions in the United States could supply advanced research in physics or engineering, the University of Akron was of the greatest importance for its intellectual and laboratory resources in rubber chemistry and its proximity to the center of the rubber industry.

University leaders knew that, while it had experts in rubber chemistry on staff, it needed to attract a specialist in synthetic rubber. President Hezzleton Simmons found that person in G. Stafford Whitby. Born in Hull, England, on May 26, 1887, Whitby had a lifelong interest in the science and chemistry of rubber.[28] At age sixteen, he won one of only five national scholarships in chemistry and used the financial aid to attend the Royal College of Science in London. With his characteristic dry wit, Whitby later said he was surprised by the prestigious award because he had fully expected not to win it until he was seventeen. After convincing his parents that he was mature enough to go to London alone at his tender age, he quickly became the top student at the school. Four years later, he graduated at the head of his class and remained with the college as a demonstrator, working with Sir William Tilden, a pioneer in the synthesis of rubberlike materials from isoprene. Despite his pioneering work, Tilden became frustrated because he could not speed up the process of making the synthetic materials and in 1907 (the year that Whitby graduated) announced that it would be commercially impossible for synthetic rubber to become a reality. He advised anyone who was interested in funding such a venture to put their money into rubber plantations.[29]

G.S. Whitby in 1907 at the Royal College of Science in London. *University of Akron Archives.*

Because Tilden believed plantations were the future of rubber, he helped Whitby earn a position in 1910 as a chemist for the Societé Financiere des Caoutchoucs at its rubber plantation in Sumatra. Although this was an important opportunity to advance his career, Whitby was at first reluctant to go. "There was a great deal said against working in the tropics in those days," he said. "I had always been frail, you see. So I thought it might have hazards."[30] Nevertheless, he did accept the position, and though he got malaria (he later described it as a mild case), the years he spent there were very productive. He became one of the first rubber chemists ever to work on a rubber plantation and used his time to research a variety of rubber problems. For example, he explored ways to package raw rubber more efficiently prior to shipment, performed vulcanization studies, and studied the physical properties of latex.[31]

Whitby on a river in Malaysia. He is the one on the right splashing in the water (1915). *University of Akron Archives.*

Whitby did find time for other interests in life. On one of his trips home to England, he became engaged to Wynne Atkinson, twelve years his senior, whom he described as someone who "radiated to the end charm, vitality and zestfulness."[32] The two first met when at the Fabian Summer School in 1909. The Fabian Society, founded in 1884, had a tradition of espousing British socialist and labor ideologies. Whitby later gave up many of his more radical economic ideas.[33] But he fondly recalled these early years of the century in London as exciting times, especially for people like him and like Wynne who were "discovering and exploring the world of the mind and the arts."[34] Whitby did not confine himself to the pursuit of science. He and Wynne were also passionate about the theater and enjoyed their interactions with literary luminaries of the day, including George Bernard Shaw and T.S. Eliot. The two married in 1915 in a judge's chamber in Singapore and returned to Sumatra to live together on the rubber plantation. They resided in a small bungalow seven miles from Kuala Lumpur.

Although Whitby enjoyed his years on the plantation, he longed to return to academic life and in 1917 went to McGill University in Canada. There he earned his masters of science in 1918 and his Ph.D. in organic chemistry in 1920. That same year, he published *Plantation Rubber and the Testing of Rubber*, which was based on his extensive field research in Sumatra.[35]After receiving his doctorate, Whitby remained at McGill and became a professor of chemistry. He spent the next decade exploring various topics such as the vulcanization reaction, the action of accelerators, and the composition and structure of Hevea rubber. His work led him to develop "ultra-accelerators," and he went on to demonstrate the roles that fatty acids and zinc oxide played in these reactions. It was also during this period that he first began to investigate the possibility of creating a synthetic rubber in the laboratory. In 1929, he left his academic post and became the director of the Chemistry Division of the National Research Council of Canada Laboratories.[36] He continued his synthetic rubber research.

Whitby sometimes interjected analogies drawn from English literature (about which he was passionate) into discussions about rubber. In a 1931 article published in *Rubber Chemistry & Technology*, he quoted Rudyard Kipling: "What do they know of England who only England know?" He then paraphrased this by stating, "What do they know of rubber who only rubber know?"[37] Whitby argued that it was a mistake for researchers interested in the physical properties of rubber to look only at rubber. To know England, one must also travel the world. To know rubber, one had to also study materials that had properties similar to rubber, thus opening a wider window through which to interpret and understand the complexities of natural rubber itself. And then, with this understanding, one could develop the elusive synthetic substitute.

Whitby remained chemistry director for about a decade before returning to England just before the start of World War II. There he became director of the Chemical Research Laboratories of the Department of Scientific and Industrial Research of Great Britain.[38] This was a high-profile position, and Whitby received letters congratulating him from numerous colleagues. A director of one company in Middlesex wrote, "My dear Whitby . . . I was delighted to read in the . . . 'Evening

Standard' recently, two paragraphs about yourself all mixed up with the nobility, gentry, dictators, kings and so on."[39] But the situation in Britain was deteriorating quickly. He sent his son and wife to Canada after Germany invaded France. Although Whitby remained through the Battle of Britain, he began looking for opportunities in the United States.

At the same time, the University of Akron's rubber chemistry efforts were in flux. Simmons was spending time in Washington, D.C. Howard Cramer had been teaching the course in rubber chemistry (and also established a new class on the "Chemistry of Latex Technology" in 1937), but in 1941 he decided to leave academia. An assistant professor at Akron for eight years, he went to work at Sharples Chemicals in Philadelphia, where he later became vice president.[40] LaVerne E. Cheney, an assistant professor in chemistry, took over teaching the rubber chemistry course for one year, but Simmons knew he needed to find a replacement who could instantly elevate the chemistry department, especially with the new importance placed on synthetic rubber.[41]

In September 1942, Simmons invited Whitby to come to the University of Akron to work on the synthetic rubber research program. This arrangement worked perfectly for Whitby, who was able to escape the dangers of war-torn Britain and research an exciting and important topic in both the United States and Canada. Furthermore, Whitby said, "It is of course a great pleasure to be re-united with my family, who have all been living over here [in Canada] while I have been at work in England."[42] He told his wife that Akron was the largest center of rubber manufacturing in the world and that it seemed like an "appropriate place in which to spend the later part of a married life that had begun on a rubber plantation."[43]

Simmons had known Whitby since the early 1920s, when the two of them presented a paper together at the American Chemical Society meeting in 1923 in New Haven, Connecticut. Later, the two co-authored a 1925 paper on the topic of vulcanization accelerators. They had found that "accelerators are capable of producing a degree of tensile strength that is unattainable in their absence."[44] While other investigators noticed this phenomenon, most attributed this to the fact that the rubber was heated for a shorter period of time when the accelerators were used. Whitby and Simmons argued that this was not the case,

Rubber laboratory at the University of Akron. From left to right, a rubber milling machine, a rubber curing press, and a tensile testing machine. *Tel-Buch* (1941). *University of Akron Archives.*

and that the accelerators themselves had a positive effect in increasing the tensile strength of the rubber. Their assessments were correct, and other researchers later discovered how organic accelerators increased resistance to heat, abrasion, oxidation, and aging. Organic accelerators "revolutionized the industry," and their importance was considered second only to the discovery of vulcanization itself.[45]

Simmons was anxious to work with Whitby again. He believed Whitby would be a perfect advisor for the government synthetic rubber program in Akron. Whitby was extremely enthusiastic about the opportunity. While he had wide experience in rubber research, he knew that synthetic rubber had the potential to change the field fundamentally. Whitby later wrote that synthetic rubber "has made rubber a field, for both laboratory researcher and the technologist, of greater and broader interest than ever before; it has quickened the pace of its scientific and technological march and expanded the scope and novelty of its future possibilities and development."[46]

When Whitby arrived in the United States, he found that the rubber shortage was affecting classroom teaching and laboratory experiments. When LaVerne Cheney taught the course on rubber chemistry in 1942, he had to have students double up on their experiments because there was simply not enough rubber to go around in the classroom. He predicted that in the next few years he would have to assign three or four students to a single experiment and have them share the results.[47]

Whitby quickly earned a reputation for himself on the Akron campus as a "quiet, scholarly, serious, but warm individual."[48] H. James Harwood, a student of Whitby's toward the end of his career, recalled how the professor walked between classes, deep in thought, staring only at the ground, unaware of passersby. He often began his lectures with a review of important recent scientific developments. A diminutive individual, he delivered each of his lectures in a loud, clear voice, his British accent always holding the attention of his class. Whitby took his teaching seriously and was not only well-prepared to discuss the topic of the day, but also took great care of his appearance, always dusting off his shoes before class began.[49]

Whitby became the first professor of rubber chemistry and director of the government-funded Rubber Research Laboratory at the University of Akron.[50] Although he was based in Akron, Whitby also worked with the

Richard H. Schmidt (left), head of the chemistry department, and Whitby monitor an experiment on a Banbury mixer, c. 1950. *University of Akron Archives.*

Canadian government as Chairman of its Synthetic Rubber Technical Advisory Committee. He served as the chairman of the Polymer Corporation, a government-financed company responsible for the construction of synthetic rubber plants in Canada. As a result, Whitby became a liaison between the synthetic rubber programs in the United States and Canada. R.P. Dinsmore, the assistant deputy rubber director in the United States, wrote Whitby, explaining that the U.S.-Canadian research exchange would be carried out "under secrecy orders"; Whitby would become the "one centralized point for dissemination of information."[51]

In Akron, Whitby's assignments were to take over the class in rubber chemistry and develop new courses, while also conducting fundamental research in rubber. He served as an important intellectual bridge to the rubber industry. He received a salary of $7,500 per year with $2,800 paid by the university and the remaining $4,700 by industry sponsors. In return, the local rubber companies were given access to and use of any commercial application resulting from Whitby's research. Whitby would assign any patentable research to the university, and the Akron rubber companies would have free use of the intellectual material. Whitby also agreed that he would not perform any private consulting work for which he would be paid separately.[52]

At his Rubber Research Laboratory at the University of Akron, Whitby directed the efforts of a few graduate students and other chemists who were working on special projects in the U.S. government's synthetic rubber program. "Through this research," he said, "we hope to contribute to the improvement of the quality of synthetic rubber."[53] His laboratory had a Banbury mixer and was also equipped to test abrasion, vulcanization, strength, flexing, tearing, rebound, and plasticity of rubber.

Whitby's laboratory was one of ten in the United States supported by the government to conduct research on synthetic rubber for the war effort. He received funding of roughly $60,000 per year for various synthetic projects.[54] The major success of Whitby's laboratory was its contribution to the development of "cold rubber," achieved in 1946 and 1947, after the war had ended; it was considered "one of the major improvements of GR-S."[55] Previously, chemists had only been able to polymerize GR-S at forty-nine or fifty degrees Celsius. Polymerizing it at lower temperature's vastly improved its physical properties—the rubber had greater flexibility, improved wearing capabilities, and better water-resistant qualities. The lower the temperature, the greater the improvement. The problem was that at the lower temperatures ice formed on the reactor, making polymerization impossible.

Whitby and his group helped to develop a "peroxamine" system that initiated butadiene-styrene emulsion copolymerization at low temperatures of only five degrees Celsius.[56] Previously, when organic peroxides were used as a catalyst to initialize polymerization, it was necessary to work at a high temperature so the peroxides decomposed quickly enough to cause polymerization within a reasonable time. To achieve this same result at a low temperature required a regent (known as an activator) that could decompose the peroxide at a lower temperature. Whitby found that polyethylene and polypropylene were most effective when the peroxide catalyst was used. This resulted in "cold rubber," which proved to be superior for automobile tires, giving them 30 percent longer tread wear than previous synthetic rubbers had.[57] The process was commercialized in 1947, and by 1949 one-half of GR-S production was converted to cold rubber.[58]

A second major research contribution by Whitby, both during and after the war, was his work on discovering the mechanism by which "popcorn" polymers formed. These were hard, insoluble polymers that appeared in styrene recovery units and gave chemists at the butadiene-styrene copolymer plants a very difficult time. Through extensive research, Whitby determined that this hardened polymer formed due to the presence of very small amounts of butadiene in the styrene. This created a mass of cross-linked polymers in which the active radicals were immobilized and trapped. Whitby's discovery paved the way toward establishing a method by which to deactivate these trapped radicals.[59]

After the war was over, Whitby focused his attention on research at the University of Akron, publishing much of the secret research on synthetic rubber he had developed during the war. Although he wrote numerous articles, his most significant effort was an edited collection entitled *Synthetic Rubber*, which represented the definitive knowledge of synthetic rubber at the time.[60] Maurice Morton, who would later take over Whitby's position at the university, called it "a capstone to an unequalled record of accomplishment in the science of rubber."[61] In 1952, Whitby turned sixty-five years old and reduced his duties at the university. The Board of Directors named him Professor Emeritus of Rubber Chemistry, and he remained as director of rubber research and continued to teach. By the end of his career, he had received every major award in his field in recognition of his importance to rubber chemistry. These included the Colwyn Gold Medal of the Institution of the Rubber Industry of Great Britain (1929), England's highest award for scientific excellence, as well as the Goodyear Medal of the American Chemical Society (1955) for, as the *New York Times* reported, "outstanding service to rubber science."[62] His final award was the Dunlop Medal of the Chemical Institute of Canada (1971). One year later, on January 10, 1972, he passed away in Delray Beach, Florida, at age eighty-five.

Even late in life, Whitby remained connected to Sumatra, where he began his career. He traveled back to the region in 1964 and in the late 1960s wrote letters to the *Christian Science Monitor* protesting the desires of the country's leaders to make Malay its official language. Always the advocate for science and technology, he believed this language was "complete[ly] unsuitable for modern, technological society."[63] But Whitby's life was more than his research, and the legacy of the man was captured in the university's tribute to him upon his retirement. It stated that he was admired "not only for his prowess as research scientist and teacher, but also for his friendly smile, his congenial manner, his wit, his amazing knowledge of any subject under discussion, and his active advocacy of the liberal education."[64] The University of Akron and the United States was made a better place by this British Renaissance man of rubber.

SYNTHETIC RUBBER ADMINISTRATION AT AKRON

Though Whitby played an important role in the government's synthetic rubber program, the University of Akron's most significant contribution to this effort was the administration of the U.S. Government Laboratories.[65] In the first decade of the synthetic rubber program (1942–1952), the government allocated over $40 million to organizations able to provide solutions to the rubber emergency. While companies like Goodrich, Goodyear, and Firestone directly received a significant portion of these funds, the University of Akron was given by far the largest amount to manage its operation of the Pilot Plant and Government Evaluation Laboratory.

INSTITUTION	GRANTS[66]
University of Akron	$7,613,613.00
Copolymer Corp. (Tire Test Fleet)	$6,083,324.00
U.S. Rubber	$4,456,982.00
Goodyear Synthetic Rubber Corp.	$3,031,369.00
Firestone	$2,985,624.00
Goodrich Chemical	$2,491,135.00
Mellon Institute	$1,447,542.00
Phillips Petroleum	$1,340,688.00
Copolymer Corp.	$1,253,461.00
National Bureau of Standards	$1,230,321.00
University of Illinois	$1,097,232.00
University of Chicago	$ 762,669.00
22 other companies (less than $750,000)	$ 750,000.00
Case Institute of Technology	$ 745,208.00
Franklin Institute	$ 691,166.00
University of Minnesota	$ 631,214.00
Cornell University	$ 597,777.00
13 other universities (under $500,000 each)	$ 500,000.00

It was surprising to some that the government gave the University of Akron this managerial authority; they argued that it would have been better invested in the National Bureau of Standards or the Rubber Reserve itself. Peter Morris, who wrote a history of the synthetic rubber program, suggested that "These more effective methods of administration may have been prevented by pressure from the rubber companies."[67] In fact, Akron's rubber company leaders, along with Simmons, were powerful advocates for the university. Ultimately, they were successful.

For the University of Akron, this was not a scientific venture. There was little interaction between the Government Laboratories it managed and Whitby's group at the Rubber Research Laboratory. The bulk of wartime scientific research in synthetic rubber was shouldered by four academic centers, which received the largest scientific synthetic rubber grants: the University of Illinois (which worked on chemical structure and organic synthesis), the University of Chicago (free radicals and emulsions), the University of Minnesota (analysis and kinetics), and Cornell University (physical methods and polymer configurations).[68] Nevertheless, the administrative role the University of Akron played was a significant one, and essential to the overall success of the synthetic rubber program.

Throughout the war years, Simmons spent a great deal of time in Washington supporting the synthetic program through his knowledge of rubber chemistry and his administrative expertise. Ironically, in 1934, Simmons was one of one hundred university presidents who sent a letter to President Roosevelt stating that war would be a "terrible disaster for the United States."[69] After Pearl Harbor, however, he devoted his significant capabilities to the cause, working with the Office of Rubber director. Created upon the recommendation of the Baruch Committee on September 17, 1942, this position concentrated on one person the responsibility of "technical research and development, importation, purchase, sale, acquisition, storage, transportation, provision of facilities, conservation, production, manufacturing, processing, marketing, distribution, and use of natural and synthetic rubber." William M. Jeffers, formerly president of the Union Pacific Railroad, became the rubber director, or as he was sometimes more informally called, the "rubber czar." Since Jeffers did not come from the rubber industry, he needed a technical team working closely with him to help the nation achieve its goals. According to the Baruch Committee, "If we fail to secure quickly a large new rubber supply, our war effort and our domestic economy will both collapse."[70] Simmons became one of Jeffers's key technical advisors, serving as the associate chief of the Rubber and Rubber Products Section of the War Production Board.[71]

Two months later, the War Production Board held a secret meeting on rubber research at the Mayflower Hotel in Akron. Attending the gathering were experts from the rubber industry and academia throughout the United States.[72] They concluded that the center of the United States synthetic rubber effort should be in Akron, and the University of Akron should play a key administrative role, as well as perform some of the research. In 1942, Simmons presented this plan to the Board of Directors at the university and also to the local rubber companies, and was met with unanimous enthusiasm.[73] The government also agreed, and in early 1943 the university entered into a contract with the Rubber Research Company to perform research in synthetic rubber.[74] During October 1943, Lee Jackson, an executive vice president of Firestone Rubber (and a University of Akron board member), and Simmons worked to promote this idea throughout industry to gain national support.[75]

In December 1943, Simmons obtained a letter of intent from the rubber director, by this time Colonel Bradley Dewey, for the university to supervise the project.[76] In February 1944, the Defense Plant Corporation built two laboratories—the Government Synthetic Rubber Pilot Plant and the Evaluation Laboratory—and also a Chemical Division. These three facilities worked on the following specific projects: the pilot plant prepared experimental synthetic rubber; the chemical division conducted research and performed control testing of the rubber from the pilot plant; and the evaluation laboratory determined the physical properties of the pilot plant rubber.[77] The University of Akron then leased the buildings from the Defense Plant Corporation and operated them under the guidance of the Rubber Reserve Company research committee.[78] The university's role, according to the agreement documents, was to manage and "conduct certain technical work and research programs in connection with the development of synthetic rubber including the operation and supervision of a Pilot Plant and Evaluation Laboratory."[79]

The laboratories were dedicated on June 28, 1944. Colonel Dewey attended the ceremonies and also provided the first opportunity for the public to tour the facilities. These tours included inspections of the buildings themselves as well as all of the rubber-testing equipment. The visitors entering the pilot plant were usually "awe-struck by the maze of pipes, conduits, and various sized . . . equipment."[80] A reception dinner was held in downtown Akron at the Mayflower Hotel, with Dewey as the guest of honor. This was Dewey's second trip to Akron in two weeks; during the first, the university conferred upon him an honorary doctorate, and he gave the address at the 1944 commencement exercises.[81]

The laboratories themselves were located on 5.8 acres of land on Wilbeth Road, several miles away from the university, but next to the Synthetic Rubber Plant operated by Firestone. Thirteen buildings covering 53,412 square feet of space comprised this complex.[82] Butadiene and styrene monomers were available to the government laboratories via a pipeline from Firestone. Steam was also purchased from Firestone, as was electricity and fire and police protection. The buildings at the government laboratories consisted of an administrative office, chemical and physical testing laboratories, the pilot plant, a building where polymers were made continuously and in batches in nonaqueous media, a drier building with equipment to dry the polymers, two storage buildings, a complete machine shop, a refrigeration building, a building that housed dangerous chemicals, and a building where large batches of chemicals were made.[83]

James W. Schade became manager of the evaluation laboratory and pilot plant. Schade's salary was $10,000 per year, which the university paid from operational funds given by the government.[84] He had worked with the University of Akron in 1929 as a member of the technical advisory committee that attempted to establish the first Institute of Rubber Research. Although this institute never materialized, Schade now had the leadership role of the government laboratory, and the university authorized him to begin hiring qualified individuals to staff its facilities. These included operators, assistants, helpers, and laborers at the pilot plant; technicians at the chemical department; and laboratory aides at the physical testing department.[85] Because of the war, it was very difficult to find civilians with the proper training to work in these laboratories. As a result, the government reassigned thirty-two chemical warfare officers to help conduct research into the synthetic rubber problem. By the end of World War II, the laboratories had a combined total of 209 employees, and the central focus was on improving the overall quality of the GR-S synthetic rubber. Schade eventually turned over operations of the pilot plant to William K. Taft, who assisted him with its growing managerial duties.

Although Whitby's group at the Rubber Research Laboratory did not have extensive interaction with the government laboratory, there is some correspondence that indicates Whitby was a resource. For example, in

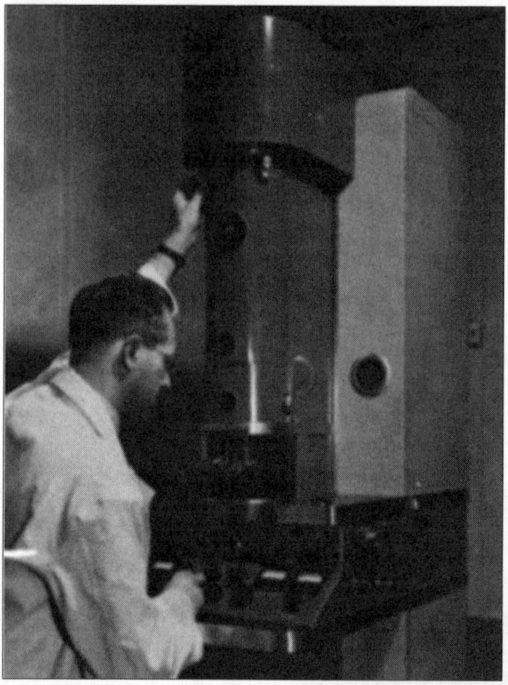

The Government Synthetic Research Laboratories operated by the University of Akron. *Tel-Buch* (1955). *University of Akron Archives.*

(*left*) A chemist studies the structure of natural and synthetic rubber using an electron microscope.

(*below*) A chemist making experimental rubber samples in the glass bottle.

December 1945, he reviewed data on a series of magnesia cures and speculated, "In some cases the data may seem out of line."[86] Whitby advocated a program for cross-checking the tensile-testing capabilities at the various laboratories. He believed the main factor influencing the results was the extent of elongation. Whitby thought that the elongation "depends on the wetting of the pigment by the polymer and is quite sensitive to conditions of mixing . . . in the vase of GR-S and magnesia."[87] Typically, Whitby used his lifelong experience in rubber chemistry to aid the government laboratory.

Investigators at the Akron laboratories were able to conduct more extensive research in the pilot plant than at operating plants like Goodyear, Goodrich, or Firestone, as they could generate rubber samples of almost any size, from small vials, to six-thousand-pound batches. Furthermore, the Akron laboratories engaged in "referee testing," a service provided to other researchers in the government's rubber programs; the government laboratories in Akron (as well as the professors at the university) worked to answer specific rubber questions from other researchers in the field. As a result, Akron became a key intellectual resource for the entire synthetic rubber effort.[88]

A large research staff working with state-of-the-art equipment made these achievements possible. The pilot plant had sixteen five-gallon reactors with mechanical stirrers and water jackets to control temperature. This equipment enabled it to produce polymers with the emulsion method in quantities from 5 to 850 pounds. Investigators developed the rubber materials in small bottles in the hopes of creating rubber with improved physical qualities. Researchers then took these experimental rubbers and subjected them to chemical and physical tests. The results helped other researchers in the evaluation laboratory select formulas that they believed would yield the best rubber in large scale production in eighty- or five-hundred-gallon containers. The equipment in the chemical laboratory included infrared and ultraviolet spectrophotometers, a polarograph, and a freezing point apparatus that could measure temperatures within a thousandth of a degree, as low as minus 150 degrees Centigrade.

In 1946, Hezzleton Simmons reflected upon the success of the government laboratory program during the war. Many new synthetic materials had been studied and great strides made in research. "The fabrication of rubbers possessing improved quality, for tire and industrial rubber goods, have been successfully accomplished," he said.[89] But just because the conflict was over, Simmons saw no reason to close these laboratories. "Even with the return of plentiful supplies of natural rubber," he argued, "a wise national policy should provide for continued production of substantial quantities of synthetic rubbers and continued research and development with a view to improving their quality and reducing their cost."[90] For the time being, the U.S. government agreed with Simmons's assessment, and the Akron Government Laboratories remained open after the war, with the university remaining as the central administrative authority.

Although the government continued to support these laboratories, in 1947 it began to cut funding. In July, the Office of Rubber Reserve reduced the budget for the Akron laboratories by 40 percent, terminated staff, and curtailed the extent of the chemical research. Despite the cuts, the laboratory made headway into new areas such as developing new rubber compounds, establishing more uniform testing procedures, and determining how to obtain the highest quality from new polymers. In 1947 alone, the laboratory produced 125,000 pounds of rubber and rubber pigment mixtures for evaluation either by industry or government. The improvement of the synthetic rubber production process was still extremely important. On January 19, 1948, the House of Representatives introduced a bill mandating that the United States should not produce less than two hundred thousand pounds of synthetic rubber per year. In March, the "Rubber Act of 1948" was passed, establishing a national policy for synthetic rubber.[91] This would help to ensure that rubber shortages would no longer threaten the national security of the United States. Simmons believed the government laboratories were essential to this process because they "are equipped both in facilities and personnel to assume an important place in any program projected by the Congress to improve the quality or reduce the cost of synthetic rubber."[92]

By 1950, the Akron laboratory saw a rise in its staff again from 200 to 220 employees, and its operating budget was $1,066,000 per year.[93] Growing concerns and tensions with the Soviet Union, significant instability in Asia with the onset of the Korean War, and a potential conflict with China, necessitated the additional funding and personnel.[94] "The Korean invasion has made it eminently plain that the entire free world lives in the constant danger of further Communist aggression," wrote President Harry S. Truman, and "The problem is particularly acute in the case of rubber."[95] The government laboratories produced some specialty polymers, which the military used on several thousand vehicles. "This work has been carried forth at a considerably accelerated rate," Simmons said, and would "continue as long as the present world situation exists."[96]

In 1951, manager James Schade reviewed the eight years during which the university had operated the laboratories and discussed several important achievements. This included road testing 100 experimental polymers, developing 350 experimental polymers for military research groups, and sending 2,500 polymer samples to various industrial and university groups. Researchers at the laboratory also wrote up the results of their own studies, producing a massive fourteen-hundred-page report that was distributed to everyone involved in the government rubber program. Schade highlighted one other key achievement, which was the discovery of a formula to make GR-S at 122 degrees Fahrenheit up to 45 percent faster than the methods currently in use by industry. B.F. Goodrich began putting this formula to use.[97] Eventually, this process meant that GR-S could be made in fifteen minutes, as opposed to the twelve or fifteen hours previously required.[98]

By 1952, the government had scaled the programs back again by about 25 percent less than the previous year's budget. It continued work on several classified projects for the Army Chemical Corps, however, shipping 13,600 pounds of experimental samples to the military, and 36,700 pounds for industry.[99] Nevertheless, the reductions indicated the eventual closure of the government rubber laboratories was imminent. Rumors began to circulate that the United States would soon start selling off these plants. This was also a time of leadership changes, and Schade, who had managed the operation since the beginning, tendered his resignation. In writing to President Norman Auburn (Simmons's successor as head of the University of Akron), Schade said, "I wish to express my appreciation of the cooperation and helpfulness the Board [of Directors at the University of Akron] has always shown in considering the problems peculiar to our operations, so different in many ways from the problems of the educational organization."[100] Some of these administrative problems included funding, ongoing labor disagreements, concerns about unionization, and efforts to establish retirement benefits and unemployment compensation. Although the university paid these workers and considered them employees, they were classified as "essential national defense workers," and as a result did not qualify for such university benefits as its School Employees Retirement System.[101] Schade retired in 1952 at age seventy, and William K. Taft, who had been managing the pilot plant, took over the entire government operation in Akron.[102]

In April 1953, President Dwight D. Eisenhower made the decision that the government no longer needed to play a central role in synthetic rubber production. "The Nation will be best served by the development within the United States of a free competitive synthetic rubber industry," Eisenhower wrote, "and I believe that now is the time to undertake plant disposal."[103] Undeterred, one month later, President Auburn conveyed to Major General E.L. Ford at the office of the chief of ordnance his suggestions for continuing the research in synthetic rubber at the Akron government laboratory. In May 1953, high-level Pentagon generals promised that some support of university research projects would continue. Auburn met with Major General Kester Hastings (Deputy Quartermaster General), Major General E.L. Ford (Army Ordnance Chief), and Lieutenant General L.E. Craigie (Air Force). During a series of conferences, he advocated $1.3 million in funding so that research projects might continue.[104]

Auburn and Whitby suggested that these projects might include military tires for use in arctic conditions, oil- and solvent-resistant polymers for temperatures below -65°F and greater than 165°F, and low temperature adhesives. Their report went on to discuss governmental funds already spent on synthetic rubber research. Through June 1953, this included $18,370,000 for industrial organizations, $7,882,000 to operate the tire test fleet, $9,958,000 for university and institute research, and $8,710,000 for the Government Laboratories at the University of Akron. The government gave Auburn no commitments, and he and Taft continued to press their case. In 1953, they met in Washington with representatives of the National Science Foundation to discuss their proposals. Auburn argued for the importance of the endeavor, stating: "Our abiding interest in this matter stems from our sincere conviction that fundamental research must be continued to prove the storehouse of knowledge so vital both to national security and industrial development."[105]

The university fought hard to retain control of the laboratories and sought ownership of them.[106] In 1954, the number of operating personnel at the Akron laboratory decreased from 176 to 138. But private industry helped to keep the work going. The laboratory's annual budget was $960,000, almost one-fifth of which came from industrial contracts, including Shell Chemical, Hercules Powder, Enjay Laboratories, Thiokol Chemical, and Standard Oil of Indiana.[107] Administrators submitted a plan to the government suggesting that it sell the laboratories to the university so they could stay open. One proposal in March 1954 was that the government sell the laboratories to the university for one dollar. Auburn had support for this plan from the local Akron rubber companies.[108] In December 1954, the presidents of these companies—Lee R. Jackson (Firestone), William O'Neil (General Tire), W.S. Richardson (B.F. Goodrich), J.P. Seiberling (Seiberling Rubber), and E.J. Thomas (Goodyear)—met in Auburn's office in Buchtel Hall to discuss the importance of transferring ownership of the government laboratories to the university.[109] The result was a letter that all of them signed to the Rubber Producing Facilities Disposal Commission recommending the sale to the university for a nominal sum. They wrote, "We believe the Laboratories should be continued not only for the purpose of serving the interests of national defense, but also to provide research and evaluation service."[110] In January 1955, the Rubber Producing Facilities Disposal Commission recommended that the university keep the laboratories open for at least one year. In managing these facilities, the commission stated that it "believes [the university] has done a very creditable job."[111]

Ultimately, efforts to keep the plants open were not successful. In December 1955, an advisory committee to the White House and Congress recommended that the government end its support of the University of Akron-managed synthetic rubber research.[112] However, the government did offer the university an option to take ownership.[113] School officials evaluated the option, but in February 1956 decided to discontinue association and control of the laboratories.[114] Harry P. Schrank, vice president of Seiberling Rubber, headed the group that "reluctantly and regretfully" made the decision.[115] With no clear stream of industrial support and with no government funding, the university no longer wanted the risk of operating the laboratories, and the relationship was officially terminated in June 1956. The advisory committee commended the school on its role in managing the laboratories, stating "The University of Akron has rendered a public service in managing and operating the laboratories for the government."[116]

The decision was met with disapproval from some in the rubber industry.[117] A *Rubber Age* editorial disagreed with the government's conclusion that the University of Akron laboratories were "no longer unique." In May 1956, *Rubber Age*'s editors wrote that when the laboratories closed, "the synthetic rubber industry as such will lose its opportunity to have a basic, fundamental laboratory devoted to monomer and polymer research, the results of which are available to the entire industry. . . . From where we sit, the Akron laboratory is in a perfect position to continue furnishing fundamental data on synthetics."[118] The U.S. government thought differently. On July 20, 1956, the House of Representatives passed a bill permitting the sale of all the government-owned synthetic rubber research laboratories in Akron.[119] Though it cost more than $2 million to construct the laboratories, the government established their current book value at $545,000.[120] In July 1957, the government put all thirteen buildings at the Akron site up for sale, and Firestone eventually purchased them.[121] The Akron era of synthetic rubber administration for the U.S. government was over.

CONCLUSION—SIMMONS'S DREAM

This did not mark the end of the University of Akron's prominence in rubber chemistry, but instead it signified the beginning of a period of change and an era of new leadership. On December 30, 1955, Hezzleton Simmons died unexpectedly at his home. Though no longer president of the school, his mind was rarely far away from his beloved

Simmons inspects the construction of a new Knight Chemical Laboratory, which was completed in 1950. *Tel-Buch* (1948). *University of Akron Archives.*

institution. He passed away listening to a University of Akron Zips basketball game on the radio. The university's Board of Directors honored him by passing a resolution stating: "He was truly a soldier on the home front as he pooled the scientific knowledge and production resources of the rubber industry for his capacity as Associate Chief of the Rubber and Rubber Products Section of the War Production Board."[122]

One of the hallmarks of the modern era is a tendency toward "creative destruction," the ever-changing process of tearing down older buildings and replacing them with structures that had newer artistic sensibilities and more functional capabilities. When institutions tear down the past, the names honored by the former buildings are often forgotten. Hezzleton Simmons was not one of these. The demolition of the old Simmons Hall in February 2000 was significant because it was the first building sacrificed to make way for the aesthetic remodeling of the campus through current President Luis Proenza's "Landscape for Learning" initiative. For a while, the name of the university's ninth president disappeared from campus, but in 2004 the new Hezzleton Simmons Hall opened as a modern center for all student-administrative needs, from financial to academic counseling.

Although Hezzleton Simmons accomplished a great deal during his tenure, teaching rubber chemistry and helping to guide the nation's synthetic rubber quest, one important dream eluded him. Simmons mentored many excellent graduate students through their master's degrees, but he was unable to take them to the highest level of academic achievement—the doctorate. Simmons believed this represented one of the major weaknesses of the University of Akron program. He often imagined the benefits of having a group of advanced graduate students working on fundamental research for the rubber industry. In 1931, he said, "With these men it would be possible to have graduate students work for their Doctor's degrees here. . . . Certainly such a scheme should some day be realized at the University of Akron."[123] His dream would not be realized during his lifetime, nor would the eventual Ph.D. program come in the old specialization of rubber chemistry, but instead, in the emerging discipline of polymers. This class of five Ph.D. students would meet for the first time, just ten months after Simmons' death.

3

POLYMERS, POLITICS, AND A PH.D.

1956–1959

IN 1938, HERMAN F. MARK AND HIS FAMILY WERE running for their lives. Mark had been a preeminent polymer chemist at IG Farben in Germany, the leading chemical company in Europe at that time, and Mark's polymer knowledge was of tremendous value to the impending war. Mark, the son of a prominent Jewish physician, was interrogated for several days at a Gestapo prison camp; the government wanted to ensure that he had shared with them all he knew about polymers. Farben would come to play an essential role during World War II, by making the synthetic rubber most Nazi vehicles rode upon, as well as fulfilling other essential wartime chemical demands.[1] Although the Nazis released Mark, they took his passport and forbade him to leave the country or have contact with anyone who was Jewish. Mark knew it was time to flee. He converted all of his money into platinum wire, hung his clothes on it, and packed all his belongings in a suitcase.[2] He dressed his family as tourists, put a Nazi flag on his automobile, and escaped to Switzerland. Mark later emigrated to the United States, where he founded the nation's first polymer research institute at Brooklyn Polytechnic in 1946. He is remembered as the "father of polymer education."[3]

By the mid-twentieth century, "polymer science" had become a new field whose practitioners explored the arrangement of large hydrocarbon molecules in long chains. This knowledge and its applications would transform the chemical industry. The development of synthetic rubber during the natural rubber emergency during World War II helped prove that polymers could become essential materials. The result was an "explosion of new opportunities," as scientists developed plastics and rubber materials that could replace metallic and natural materials of all kinds.[4] The emergence of a true science of polymers required a dedicated intellectual effort in basic research extending beyond the capabilities and the commercial sensibilities of industry alone—it needed an academic discipline and advanced training at the university level.

Although Mark had founded the first polymer institute, his students earned their Ph.D.'s in physics or chemistry with a specialization in polymers. At the time, no institution offered a distinct polymer chemistry program granting a doctoral degree. This would change thanks to two men at the University of Akron—President Norman P. Auburn and polymer chemistry professor Maurice Morton. They had both succeeded giants at the university, with Auburn becoming president after Hezzleton Simmons in 1951 and Morton taking over from G. Stafford Whitby in 1952. Both quickly left their indelible marks on the university, establishing the Institute of Rubber Research (March 1956) and one of the first Ph.D. programs in polymer chemistry in the world (September 1956).[5]

While rubber research had become a vital national issue during Simmons's tenure (due to World War II), political events also played a role in justifying the Akron polymer program in the 1950s. Both Morton and Auburn were very much aware of the past significance of rubber to national security.[6] But new concerns emerged in the postwar years. Rubber, polymer, and university education took on greater political significance with the growing threat posed by the Soviet Union. Both men looked toward advanced graduate training opportunities and fundamental polymer research as important ways to strengthen the United States for future military or ideological conflicts. In an era of space and arms races, some began calling this the "education race." Auburn, who had taken several fact-finding trips throughout the Soviet Union, said, "An important lesson to be learned from the Sputniks is that America has failed to support the various elements of higher education to the degree required. That must now be rectified."[7] He believed that Sputnik represented the end of one era and the beginning of another.[8] Only through education, or as Auburn called it, an "E-bomb," could the United States compete and survive.[9] Morton referred to it as the "rubber race." A doctoral program in polymer chemistry could not only offer one path toward winning the education race, but the fundamental basic research produced could potentially be a key component of future military efforts.[10] Morton and Auburn's efforts culminated in an advanced polymer program that would rival, if not surpass, the contributions made by the university in the field of rubber chemistry. It was an era in which polymers, politics, and Ph.D.s were intimately connected.

FOUNDING POLYMER SCIENCE

The emergence of polymer science is considered by many to be one of the scientific milestones of the twentieth century. Historian Yasu Furukawa concluded, "The birth of polymer chemistry . . . marks a new epoch in the history of modern science."[11] In his masterful *Inventing Polymer Science*, Furukawa observed that, in science, theory formation and discipline formation are often two separate processes.[12] Polymer science followed such separate paths. In a sense, the discipline remained unformed while scientists established not just the underlying theories, but even the name of the pursuit itself. The knowledge of "polymers" existed in the nineteenth century; however, the word's original meaning was different from our current definition. "Polymers" described any molecule that had repeated atomic units, regardless of size. Jöns Jacob Berzelius, a Swedish chemist, initially coined the term in 1832. He devised the word "polymeric" to refer to substances that had similar numbers of atoms, with the same composition, but different properties.[13] It would be almost a century until modern "polymer science" was born.

The actual birth of a polymer science occurred in the 1920s when researchers recognized that polymers were actually very large molecules, consisting of chains of atoms, linked through chemical bonds. The years immediately following this initial breakthrough were marked by significant scientific discovery and intense intellectual, and sometimes personal, debate and disagreement. As one observer said, the development of polymer science "does not present a picture of peaceful idylls."[14]

Hermann Staudinger laid the groundwork when he introduced the concept of the "macromolecule" in 1922. Staudinger was born in Worms, Germany, in March 1881, and after an early interest in botany studied classical organic chemistry, earning his doctorate at the University of Halle in 1903. His research soon established him as an important theorist, and in 1912 he was appointed professor at the Eidgenössische Technische Hochschule (ETH) in Zurich. The ETH was known as one of the most influential technical schools in Europe, a reputation enhanced by the success of one of its graduates, Albert Einstein, in 1900. (Einstein returned to the ETH to become a professor of physics the same year that Staudinger entered the chemistry department.) Staudinger quickly established his own reputation as students soon "came under his spell" when he "penetrated into the mysteries of the carbon atom."[15]

It was at the ETH, during World War I, that Staudinger first conceived of a theory of polymers.[16] Staudinger represented the "organic-structural tradition" in chemistry, believing that polymers could best be understood by interpreting them from the internal structure of the organic molecule.[17] Colleagues initially dissuaded him from studying polymers at all, believing it was an area that consisted of "poorly defined compounds" like rubber and was "very unpleasant."[18] Staudinger paid no attention to their warnings and ultimately helped to found a new science. In 1922, he demonstrated his first evidence for the existence of a macromolecule and the long chain of large molecules. In an experiment on the hydrogenation of rubber, he suggested that the particles of rubber were not held together by small particles, but instead by large, linked ones. He intentionally used the term "macromolecule" to avoid the name "polymer" and its nineteenth-century connotation of small repeating molecular substances.[19]

The field today is not typically called "macromolecular science" due to a controversy between Staudinger and his contemporary, Herman Mark. Not only did Staudinger and Mark have somewhat differing views on this emerging science, but they also had deeply contrasting personalities. Staudinger was an imposing six-foot-four-inch German, with nervous mannerisms, a quiet voice, and a low tolerance for different opinions. He often frightened his students, who sometimes hurried from his laboratories when they heard him coming. Even G. Stafford Whitby, a renowned rubber chemist himself, once said that Staudinger was "not exactly [a] gentle and compromising protagonist" of his own work.[20] In contrast, Mark had a warm, rosy face and, according to one *New York Times* reporter, would sit without a "muscle trembling" behind his laboratory desk, beneath a leaping sailfish that symbolized his passion for the outdoors. His visitors were often greeted with Beethoven symphonies, and Mark always offered them his "warm concentration."[21] Mark, of course, had fled Nazi persecution; Staudinger neither fled nor supported the Nazi regime, but his chemical laboratories were nearly destroyed in 1944 during the air raid on Freiburg.[22] These two men engaged in a bitter controversy over who was the first to accurately define the nature of the long-chain molecule. Mark would never use the macromolecule terminology and instead became the "peripatetic chief missionary" for what he called "high polymeric chemistry" in the 1930s.[23]

Mark was born in 1885 in Vienna. His father, a physician, often came into contact with other prominent professionals, including Sigmund Freud. In 1913, Mark joined the Imperial Austrian Army and served throughout World War I. After four and a half years of trench warfare and more than fifteen medals for bravery, he was captured at the Italian front and held as a prisoner of war for eight months. After the war, he studied organic chemistry at the University of Vienna, earning his doctorate in 1921. In the pioneering research he did while working at IG Farben, Mark used x-ray analysis to discover for the first time the structure of cellulose and other artificial polymers. At the time, Farben dominated the German chemical industry, and from 1925 to 1945 it was the largest private corporation in Europe.[24] Company leaders demanded a high level of academic excellence from its researchers. Mark's research group was encouraged to publish scientific papers and attend conferences just as their university counterparts might. Mark managed a group of more than fifty scientists, and his work enabled Farben to manufacture its first commercial polystyrene in 1930.[25] But Mark and his associates were at the same time advancing a new polymer theory which, though in agreement with Staudinger's long-chain molecules, differed in several ways.

While Staudinger came from the "organic-structural" tradition, Mark viewed polymers from the "physicalist" tradition. Staudinger believed that molecules and high, uniform molecular weights were central to macromolecular science. Mark did not disagree with the high molecular weight but wanted to explain crystalline and noncrystalline regions and a nonuniform range of molecular weights. He and his research associates believed that polymers were, in fact, aggregates of various-sized molecules bonded together by physical forces.[26] Mark and his group at Farben called them "micelles," aggregates of primary valence chains. In 1930, Mark advanced these ideas by writing the first textbook using his proof of the macromolecular concept. However, to differentiate his theories from Staudinger's, he used the term "polymers."

Staudinger was furious about the book. He attacked Mark's group for plagiarizing his work, being unscientific, and disseminating mere propaganda. Staudinger published his definitive textbook two years later, but Mark had won much of the momentum in the growing field. Staudinger frequently returned from scientific meetings in deep consternation, and his research assistants often had to calm him after his increasingly antagonistic confrontations with students and colleagues. The debate created constant tension in Staudinger's laboratory, and his graduate students formed a strong bond that resembled, as one student recalled, "a cult of believers who would sacrifice themselves for a religion."[27] Staudinger would not be recognized for much of his pioneering work until the 1950s.

The disagreements between these two camps continued until an American, Wallace Carothers of DuPont, embarked on a series of experiments that eventually established modern polymer theory. Born in Burlington, Iowa, in April 1896, Carothers had, from a very early age, a passion for tools and experimentation. But his first field of study was in accounting at a commercial college in 1915. After graduating, he entered a small liberal arts school, Tarkio College, in Missouri, where he began taking chemistry classes. He attempted to join the army during World War I but was rejected for physical reasons, most likely a goiter. Carothers continued his work in chemistry, eventually earning his Ph.D. in organic chemistry in 1924 at the University of Illinois. His instructors considered him one of the most brilliant students they had ever taught. Two years later, Harvard University offered him a position as an instructor in organic chemistry. Carothers accepted, but he remained there for only one year, until DuPont invited him to head its new fundamental research program in organic chemistry.[28] In 1929, he reflected on the uncertainties in this emerging field: "Whatever the term polymer may mean now, it does not mean precisely what Berzelius intended."[29] Carothers's work at DuPont would help to establish finally the foundations of polymer science.

In 1927, DuPont's chemical director, Charles Stine, who had available DuPont's massive financial, intellectual, and laboratory facilities, initiated a program of fundamental research into new areas of basic science. He targeted polymers as one important field of study.[30] Stine hired Carothers to head up this research, and between 1929 and 1931 he created the groundwork for organic polymer science. Carothers studied condensation polymers, an area Staudinger neglected. Condensation polymers include polyesters, polyamides, and polyacetals—they are any type of polymer created through a condensing reaction that releases a small molecule such as methanol or water.

Carothers's theoretical pursuits led him to make artificial macromolecules at a time when some scientists questioned whether or not this was even possible.[31] In 1930, one of his experiments involving acetylene solidified in the test tube into something unexpected. Carothers reported that it was "something that had never been seen before in this work—a ball with a lively bounce and other characteristic physical properties of natural rubber."[32] It was a synthetic rubber, later known as Neoprene. Carothers also discovered how to produce nylon. This would become the first manmade fiber with properties comparable to materials previously found only in nature. Polymers were becoming a "science of manmade materials" that included synthetic rubber, which is what made the entire field so essential to overcoming the crisis of the natural rubber shortage during World War II.[33]

Through his dramatic experimental successes, Carothers was able to put to rest much of the theoretical debate generated by Staudinger and Mark. In 1935, the Faraday Society in London organized a symposium at the University of Cambridge, the first international gathering devoted to polymers. For the first time, Carothers met Staudinger and other luminaries in the field. Mark also attended, and described the conference as one of the "milestones of modern chemical history." There was no longer any question about the existence of macromolecules. Staudinger had to concede the importance of condensation polymers, which he had earlier dismissed, but was gratified in that Mark's micelle structure was also abandoned in favor of the supremacy of the macromolecule. "At the end of the symposium," Mark said, "everybody was convinced that polymer chemistry had grown into a full-scale science with unexpected new vistas. . . ."[34]

Despite his dramatic professional success, Carothers led a troubled life. He suffered from depression and believed he was a failure. After the tragic death of his sister and fearful that he had no further ideas to contribute, he concluded that his life was not worth living. In April 1937, he took his life in a Philadelphia hotel room, swallowing a potassium cyanide pill. Guests from nearby rooms heard "groans" coming from behind his door, and when the hotel manager forced his way in, he found a few grains of the poison near Carothers's body, along with a squeezed lemon. He had apparently ingested the poison with lemon juice. The *New York Times* reported that the "co-inventor of a synthetic rubber" was found dead.[35] His wife was two months pregnant with their only child. Two years later, DuPont began producing nylon, and because of Carothers's contributions the polymer industry was born.

Carothers's legacy lived on, as colleges and universities throughout the world slowly began offering academic courses in polymers.

It was Mark who emerged as the "father of polymer education" and founded the first polymer institute. Mark did not remain in Europe long after the Faraday Society conference. He and his family had initially fled Germany for Austria in 1932, and in 1938, following his Nazi interrogation, they escaped into Switzerland.[36] Mark and his family eventually arrived in the United States in 1940, where DuPont acquired his services. DuPont sought out European refugees with polymer backgrounds like Mark's to continue the work started by Carothers.[37] DuPont offered him a joint position as consultant to the company and adjunct professor at Brooklyn Polytechnic—a position that DuPont funded. Two years later, Mark founded the Polymer Research Institute of Brooklyn Polytechnic.[38] In 1946, he established the first English-speaking journal devoted to the subject and called it the *Journal of Polymer Science*. G. Stafford Whitby, from the University of Akron, served on the advisory board. The "polymer" terminology for the profession finally became standardized, over Staudinger's preference, the "macromolecule." However, both terms are in use today.

Eventually, Staudinger received credit for his own pioneering accomplishments in the field. Although for decades his contributions were ignored, in 1953 he was awarded the Nobel Prize for his work on macromolecules, which led to the modern plastics industry.[39] He became the twentieth German to win the prize.[40] Mark, later in his life, tried to overcome his early disagreements with Staudinger and wrote Staudinger's obituary after his death in 1965, as well as a tribute entitled "Hermann Staudinger—Father of Modern Polymer Science."[41] Although Staudinger would have appreciated the high praise, he likely would have recoiled at the title of the tribute.

After the war, the knowledge of polymers began to spread beyond scientific circles. The first usage of the term "polymer science" in the *New York Times* appeared in a 1947 article announcing a chemical award presented to Paul Flory for his work on "high polymers" and "polymer science."[42] Flory was another "towering giant" in the polymer field, with ties to DuPont.[43] After graduating with his doctorate in 1934 from Ohio State University, he went to work at DuPont, where

Carothers served as his mentor. Flory was devastated by Carothers's suicide. Eventually, he left the company to research polymers at Goodyear during World War II, and worked with a group that helped develop synthetic rubber. After the war, his career alternated between industry and academia at the University of Cincinnati, Standard Oil, Cornell University, and Stanford University. In 1953, he published his landmark *Principles of Polymer Chemistry*.[44] In 1974, he won the Nobel Prize for his research comparing polymers after dissolving them in a particular solvent. He also discovered a "universal constant" by determining a polymer's viscosity, light dispersion, and sedimentation rate.[45] Throughout his career, Flory had an important relationship with the University of Akron.

Discipline formation in polymer science began after World War II. Prior to the war, there was essentially no higher education that focused directly on polymer chemistry.[46] After 1945, colleges and universities started offering systematized programs of study leading to degrees in the field. Although Mark's Polymer Research Institute was the first, a few others immediately followed. By 1956 several American universities were offering coursework in polymers, rubber, and plastics.[47] These schools taught a total of fifty-eight different rubber and polymer courses. The University of Akron offered the most courses anywhere in the country (nine), followed by Case Institute of Technology (seven), and the University of Southern California (six). The polymer terminology had become standardized, and none of the course titles mentioned macromolecules.

What the discipline of polymer science needed was for more of these schools to offer graduate degree programs and not just sporadic coursework.[48] Through the pioneering efforts of Maurice Morton and Norman P. Auburn, the University of Akron was able to begin its Ph.D. program in polymer chemistry in 1956—a milestone year not just for this school, but also for the entire discipline.

PREPARING FOR A MILESTONE YEAR

The public stereotype of the scientist at work is often a quiet, reserved scholar, more at home with laboratory devices than with people.

Maurice Morton was nothing of the sort. Robert Pierson, of Goodyear once said that Morton was endowed with a "triple helping of bounce, humor, and enthusiasm."[49] In 1956, one reporter described Morton as a vibrant ex-Canadian "who speaks of rubber with an intensity most men reserve for Marilyn Monroe."[50] In a similar story, Morton used to recall that when, early in his career, he discussed "polystyrene" energetically, and some uninformed listeners thought he was talking about a woman anyway.[51] Morton channeled his enthusiasm into a vigorous career as a researcher in polymer science and as a teacher who established close, lifelong bonds with his students.

Morton was born in Latvia in 1913. His father worked at Singer Sewing Machine Company, where he excelled in sales to rural families who before had mended and made clothing solely by hand. After World War I, political instability and change threatened the safety of the family. Some of Morton's earliest memories were of the frightening sounds of gunfire when the Russian Revolution began in St. Petersburg.[52] Morton had watched in awe as wealthy Americans drove the only automobiles in Latvia, graciously handing out chocolate to children. He began dreaming of a better life in another part of the world. He and his family emigrated to Canada, settling near relatives in the Montreal area, where Morton attended McGill University.[53] While a student, he met and married Lilian Rosenbloom, beginning a sixty-year union. He earned his bachelor's degree in chemistry in 1934 and then worked as a chief chemist at Johns Manville in Quebec (1936–1941) and at Congoleum Canada in Montreal (1941–1944).

Not satisfied with a career in industry, Morton, while still working at Congoleum, returned to McGill to begin graduate school. His early research was in the emulsion copolymerization of butadiene-styrene (which was known as GR-S). A lone wolf in the polymer field, he was most likely the only student in Canada working on emulsion polymerization. At the time he did not really know much about rubber, and in an attempt to become more informed, he decided to attend a lecture at a Canadian technical meeting. He recalled his disappointment when the term "polymer" was never mentioned. In retrospect, he said, this should not have been surprising, because all of the researchers at the time were busy " 'curing' rubber, not making it."[54] His research eventually led to a Ph.D. in 1945 and his first steps toward a reputation as a pioneer in the field. After graduation, he became the head of the chemistry department at George Williams College (now Concordia University) in Montreal (1945–1948), while also supervising graduate students at McGill.

In 1948, G. Stafford Whitby contacted Morton and invited him to the University of Akron to become his assistant. Whitby had served as Morton's outside examiner for his Ph.D. examinations, so he knew Morton's intellectual abilities well. Morton believed this was a wonderful opportunity "to help usher rubber technology . . . into the wonderful world of polymer science."[55] The presence in Akron of Paul Flory, whose work Morton respected, also attracted him to the area. The two soon became friends. This relationship was strengthened further when Morton designed experiments to confirm several of Flory's polymer theories.[56] Morton became the assistant director of rubber research at Akron and joined a staff of five other chemists who were researching various problems for the U.S. Government Synthetic Rubber Program.

The most significant new polymer initiative at the University of Akron around 1947 was the inauguration of the Polymer Lecture Group program. This became an ongoing series sponsored by the university and the Rubber Division of the American Chemical Society, featuring lectures by academic and industrial authorities in "high polymer science."[57] The advisory committee for this lecture series represented a remarkable grouping of industry leaders. Paul Flory, of Goodyear, was its chairman, and Whitby served as vice chairman. R.J. Reid (Goodyear) was the secretary, H. Leaderman (Firestone) was the program chairman, and C.F. Gibbs (Goodrich) was the treasurer. The first year, the university sponsored seven speakers.

The lectures took place in Simmons Hall (named for former president and professor of rubber chemistry Hezzleton Simmons) on the university campus. Students could attend all seven lectures for a total of $3.50; for anyone else, the cost was $7. The participants the first year included a variety of leading scholars in the polymer field: P. Debye of Cornell University presented a lecture on "Light Scattering and Viscosity"; Paul D. Bartlett of Harvard University presented "Initiators,

The Chemistry Department at the University of Akron in 1955. Seated from left to right are Maurice Morton, Thomas Sumner (department head), and Vaughn Floutz. Standing from left to right are Howard Stephens, Alvin Wolfe, and Walter Cook (Gerald Corsaro not pictured). *Tel-Buch* (1955). *University of Akron Archives.*

Retarders, and Inhibitors in Vinyl Polymerization"; Frank R. Mayo of United States Rubber presented "Copolymerization"; Walter H. Stockmayer of the Massachusetts Institute of Technology presented "Molecular Weight Distribution"; Harold M. Spurlin of Hercules Power presented "Solution Properties of Cellulose and its Derivatives"; Raymond M. Fuoss of Yale University presented "Electrical Properties of Polymers"; and Raymond F. Boyer of Dow Chemical presented "Plasticization and Second Order Transition."[58]

While the Polymer Lecture Group was bringing the nation's polymer leaders to campus, university leaders were planning an important improvement to the chemistry department's facilities. One year after Morton's arrival, in 1949, the board of directors began making plans for a new chemistry building. The original Knight Laboratory was now nearly forty years old, and the program's growth required a newer, more modern space.[59] Walter Cook, head of the chemistry department, assisted with the designs, and Lee Jackson, the chairman of the

Knight Hall in 1956. *Tel-Buch* (1956). *University of Akron Archives.*

building committee (and president of Firestone), offered the services of his company's engineering department.[60] The new building, also called Knight Chemical Laboratory, was opened to the public at a dedication ceremony on Founder's Day in January 1951. Students had already begun taking classes in it during the fall 1950 semester, and the laboratories had opened in October.[61]

The chemistry community hailed the new building as "one of the most modern scientific laboratories to be built on any university campus."[62] The forty-four thousand-square-foot building cost $1 million and took fifteen months to build. The building housed eight staff laboratories, eleven student laboratories, and space for other specialized equipment, including rubber mills, presses, physical testing equipment, and chemical storerooms. The Rubber Research Laboratory shared the building with the larger chemistry program, including organic and inorganic chemistry, and the various laboratories associated with these pursuits. The chemistry professors believed these new facilities would make the University of Akron a "true center of science."

G. Stafford Whitby retired one year after the dedication of the Knight Chemical Laboratory. He had reached the mandatory retirement age of sixty-five, and the university made him professor emeritus of rubber chemistry in September 1952.[63] The board of directors appointed Maurice Morton as the first professor of polymer chemistry and also made him assistant director of the chemistry department. This new "polymer" position, school officials believed, was "in keeping with our policy of strengthening the instructional and research work in rubber and allied fields."[64] At the time, Howard L. Stephens was an instructor in rubber chemistry, and he and Morton comprised the total full-time polymer chemistry faculty. The two offered almost all of their courses in the evenings and on Saturdays, to accommodate those graduate students who worked during the day.[65]

While the new Knight building was under construction, university administrators were in the midst of another decision essential to the institution's future. In 1950, President Hezzleton Simmons had also reached the mandatory retirement age, and the school struggled to find a successor. The search lasted for over a year and included serious interviews with fifty leading candidates. The board of directors finally chose Norman P. Auburn, a forty-six-year-old educator who had been the vice president and dean of administration at the University of Cincinnati.[66]

What Auburn found when he came to Akron contrasted with the celebrations that surrounded the Knight dedication. One university dean wrote him a letter describing the situation: "The morale of the faculty has been low indeed . . . we are a melancholy group."[67] Significant problems afflicted the university, including low enrollment, decreased income, faculty cutbacks, and reduced salaries. The entire school had an annual budget deficit of $150,000, most of it due to a lack of student tuition. To Auburn fell the demanding task of remaking the university, establishing new ideas, and revitalizing morale.[68] He brought with him an understanding of the importance of an organized scientific research program, which he and others believed might help to turn the entire university around.[69] With Auburn at the helm, Morton in charge of polymer chemistry, and the entire chemistry department now in its new building, the university was at the

Dr. Norman P. Auburn became president of the University of Akron in 1951. *Tel-Buch* (1952). *University of Akron Archives.*

brink of a new beginning. As Auburn and Morton prepared to create a new rubber institute and a doctoral program in polymer chemistry, an unlikely development served to justify and reinforce the need for these programs—the Soviet Union and the scientific race during the Cold War.

THE "RUBBER RACE"

In a 1956 television address announcing his run for reelection, President Dwight Eisenhower said, "We witness today . . . the rise of the first great industrial power to challenge the West."[70] Eisenhower argued that the Soviet's scientific and engineering capabilities represented a growing threat to the United States. He feared that the current numbers and abilities of American scientists and engineers might not be sufficient to match Russian research accomplishments. The launch of Sputnik one year later highlighted these concerns. Many observers believed this Russian satellite traveling in the nighttime sky for all to see demonstrated the disastrous consequences of the strangulation of research in the United States.[71] One U.S. governor ridiculed American science and technology:

> Oh little Sputnik
> With made-in-Moscow beep
> You tell the world it's a Commie sky
> And Uncle Sam's asleep.[72]

Both Maurice Morton and Norman P. Auburn were acutely aware of political and scientific developments in the Soviet Union. They were concerned with the growth of communism on the worldwide stage. Renewed efforts at the university level to advance scientific training and research would, they believed, go a long way to counteract and neutralize this political threat.[73]

In linking science to the political context of the Cold War and the security of the nation they were not alone. There was some real cause for concern. In the 1940s and early 1950s, Stalin had terrorized Russian scientists by condemning them into gulags.[74] This situation changed dramatically in 1953 when Nikita Khrushchev gained power, denounced Stalin, and rehabilitated the political and scientific outcasts. He elevated scientists as the country's new elite, and thousands emerged from the prison camps. Party leaders began arguing that scientific and technical experts would propel the communists into world dominance. As the twenty-second Party Congress later stated, the ultimate goal of the

Soviet Union was to "occupy *the leading place in world science* in all the key fields."[75]

Science in the United States was becoming increasingly important for national security, and scientists emerged from World War II as heroes capable both of unleashing destructive powers of the atom and developing synthetic versions of natural rubber. One journalist noted that the "lesson of World War II . . . [was that] the outcome of war was now decided, as much as anything, by a nation's scientific and engineering wizards."[76] Their prominence grew during the Cold War, and most scientists became actively aware of their political influence for the first time. Reports of Soviet technical and scientific superiority in the 1950s reinforced the need for action by U.S. scientists and engineers.[77]

Norman Auburn took a personal interest in monitoring Soviet progress. In 1956, he and his wife visited Russia, spending two weeks behind the Iron Curtain.[78] This was his second trip to the region, and upon his return, he gave numerous talks recounting what he saw and warning his listeners about what it meant for the future of the United States. In 1956, he completed his term as president of the Association of Urban Universities, and at its annual meeting he gave a long presentation about his trip. The Soviet emphasis on technical education, he said, was a primary concern.[79] The United States was falling behind the Soviets in the growth rate and quality of scientific and technical students. Auburn quoted Alfred North Whitehead's rule for modern life: "The race which does not value trained intelligence is doomed."[80] One way to counter this threat and win the race was educational and research institutes like one proposed for rubber at the University of Akron. Auburn said that the Institute of Rubber Research would play a role in ensuring the nation did "not fall behind in the technological race."[81] Auburn believed rubber research and advanced doctoral training at the university would play a vital role in the success of future military projects. Education would help the United States emerge victorious from the Cold War. Auburn's views resonated with Maurice Morton, who still recalled vividly his childhood fears of the Russian Revolution. Morton predicted that the United States would become engaged in a "race in scientific problems in rubber which are vital to the success of such projects as jet-engines, missiles and rockets, and satellites."[82] The "rubber race" was born.

After Auburn's 1956 trip to Russia, the Alfred P. Sloan Foundation in New York City awarded him a grant to continue his research about Soviet scientific and technical training. He used these funds to make another trip in 1959, when he met with the Soviet minister of higher education.[83] He spent five days in Moscow and then traveled to Leningrad and Kiev to visit the Soviet Academy of Sciences. In Kiev, he also toured a rubber research institute.[84] At the Soviet Exhibition of Science, Technology, and Culture, he saw a sign claiming that Russia was the first nation to produce synthetic rubber. Auburn also visited what he called a "must stop"—the Kautchouk Rubber Works in Moscow. This factory employed over six thousand workers and produced industrial rubber goods. Upon his return, Auburn delivered an even more dire message. He believed the Soviets were winning the race against the United States without a military engagement. To combat their progress, Auburn said, "Education is our weapon."[85] Auburn called his plan an "E-Bomb." It would cultivate basic research and discover new knowledge to keep pace with Soviet advances.[86] By improving the U.S. capacity for producing scientists, Auburn believed the Soviet scientific challenge could be countered and neutralized. By "nourishing" the educational institutions that trained the experts, Auburn said, new graduates would "carry on the scientific research to keep the nation strong."[87] Morton and Auburn both viewed a Ph.D. in polymer chemistry as one such educational weapon and used the political context of the Cold War as an important justification for it.[88]

The Cold War lasted for several decades and, ironically, by the end of it, one prominent polymer scientist offered to risk his own life to assist a Soviet scientist. In 1984, Soviet dissident and Nobel physicist Andrei D. Sakharov pleaded with the Soviet government to let his wife travel to the West for a heart bypass operation.[89] Paul Flory was one of a group of scientists who essentially offered themselves up as hostages to stand in for Sakharov's wife while she sought treatment. Flory said this would ensure that she would not take the opportunity to criticize the Soviet government or defect. Although the Soviets refused his offer, Flory continued to speak out against human rights abuses, saying "We all are scientists who have been working to realize the unity of science around the world."[90] Flory went on to help others trapped by communist governments. One example was Virgil Percec, who defected from Romania in 1981, leaving his family behind. To help bring Percec's family to the United States, Flory used a keynote address he was scheduled to give at a conference in Romania as leverage, threatening to criticize the Romanian government severely in the address if it did not let the Percec family go. The Romanian government capitulated, and Virgil Percec eventually became the Leonard Case Professor of Macromolecular Science at Case Western Reserve University.[91]

This idea of a unified, worldwide, scientific community did not exist at the start of the Cold War. In the 1950s, the focus was not on how to save Soviet scientists, but instead on how to compete against and outthink them. Morton believed his university should not only continue to take a leading role in rubber research, but also to serve as an example in producing a new level of highly trained doctoral students.[92] He argued, "There is no doubt that we are in a life and death competition with the Russians" for the scientific brainpower and knowledge that could provide the winning edge in warfare. Morton's solution, which would swing the balance back in favor of the United States was a renewed focus on the graduate student. "What can be done and when?" he asked, answering the "final stage of the [educational] process, with the graduate student."[93] Because of Morton and Auburn, the Institute of Rubber Research and also the Ph.D. program in polymer chemistry emerged within the political context of the Cold War as one way to help the United States win the "rubber race."

ESTABLISHING THE INSTITUTE OF RUBBER RESEARCH

Through the early 1950s, the University of Akron's Rubber Research Laboratory maintained its focus on basic research. Its funding came from a variety of outside sources. In 1955, these included the National Science Foundation ($50,000) for fundamental investigations of diene polymerization; the Signal Corps ($48,000) for degradation studies on condensation polymers and also the study of balloon lattices; the U.S. Air Force ($19,000) for branching in high polymers; and the U.S.

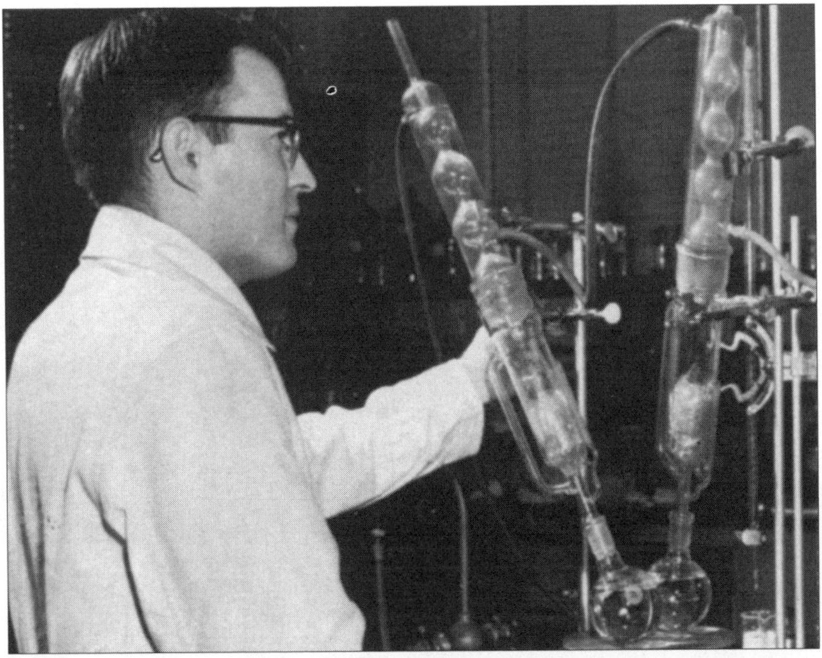

Chemistry department laboratories in 1955. *Tel-Buch* (1955). *University of Akron Archives.*

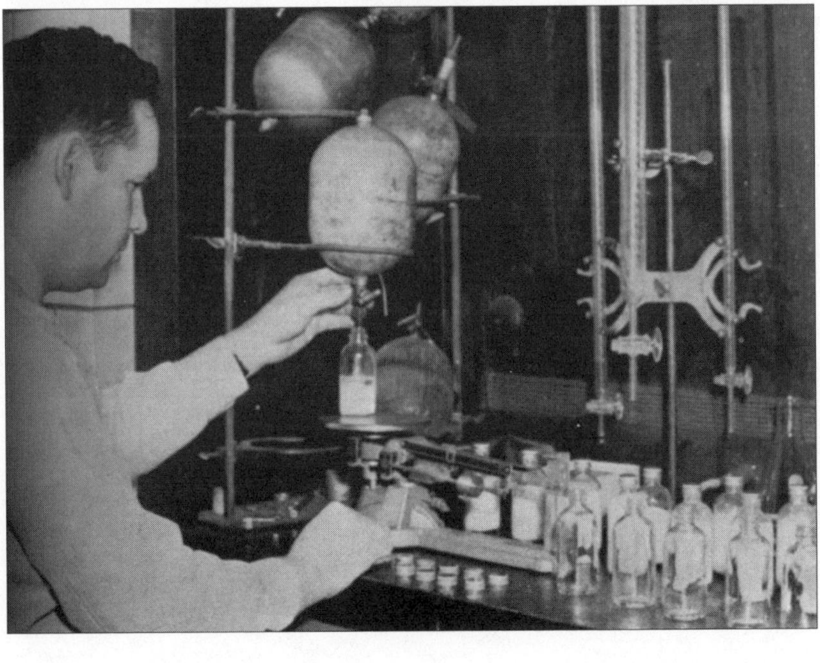

Quartermaster Corps ($18,000). Also, for the first time, the Rubber Research Laboratory obtained a research contract from private industry when Goodyear funded a $12,500 research project on fundamental studies of vulcanizate structure.[94] That private industry contract indicated that research was becoming more important for the local rubber companies, and that the Rubber Research Laboratory could evolve into a stronger and more autonomous entity within the university.

By the mid-1950s, there were twelve researchers working in the laboratory. Six of these were full-time chemists, including Maurice Morton, Thomas Sumner (the chemistry department head), Vaughn Floutz, Howard Stephens, Alvin Wolf, and Walter Cook. Three others J.P Berry from England, Michel Montu from France, and Masamoto Watanabe from Japan, were working on postdoctoral research, while the final three were Master's students. Other research chemists in the program offered assistance on the contract projects. For example, in 1955 Irja Piirma split her time between the Signal Corps and the air force projects.[95] The staff published six scientific papers that year, and graduate students in chemistry were winning national awards, including those offered by *Rubber Age* for outstanding theses.[96]

Despite these successes, Morton believed he could improve on this structure in a broader sense. He lamented that there were few basic research institutes in the United States like those at Stanford or Carnegie Mellon. Akron was a prime candidate for an institute that could unite the rubber industry with the school's rubber research. Morton believed the previous Rubber Research Laboratory's goals were too "modest," and so he began exploring ways to establish a larger and more autonomous infrastructure for its activities.[97] In 1955, he wrote, "It is recommended that serious consideration be given to the possibility of stabilizing this activity," into what he initially called the "Rubber Research Institute."[98] In January 1956, Ernest H. Cherrington, Jr., the dean of Buchtel College's liberal arts program and director of graduate studies, formed a committee to explore the project's feasibility. Morton hoped the committee would approve recommendations so that he could begin implementation in the spring of 1956. Both Morton and Thomas Sumner were members of this committee, along with representatives from all of the university colleges.[99]

Thomas Sumner, the head of the chemistry department, was a driven scientist and a prodigy. He was born April 16, 1926, and received his Ph.D. from Yale University when he was just twenty years old. He taught in the chemistry department there for four years before coming to the University of Akron in 1950. Sumner took over as head of the chemistry department in 1953 (from Walter Cook), serving until 1961.[100] He made national headlines when, in the late 1950s, he produced groundbreaking chemistry films for undergraduate students. The Fund for the Advancement of Education provided a $23,000 grant, which allowed him to create more than twenty films.[101] The films focused on laboratory experiments, with close-ups clearly showing how chemical reactions caused changes in temperature, weight, shape, and color.[102] For example, in the forty-six-minute film *Indicators of pH* Sumner showed how to prepare several different solutions of hydrochloric acid. Sumner concluded by determining the pH of an acetic acid solution and then calculating its ionization constant.[103] He was convinced that films like these would save laboratory time, space, and expenses. One journalist from *The Wall Street Journal* reported that Sumner and his film project was a "trail-blazer on the college level."[104]

Morton and Sumner did not always see eye to eye. Both were strong-willed and brilliant, and sometimes their passions for the school resulted in a contentious interpersonal relationship. Morton preferred to work closely with Auburn, and together they collaborated to develop a plan for the new rubber institute. Morton initially wrote to Auburn that he "would welcome a discussion . . . concerning rubber research," and he began by laying out his educational vision.[105] Morton first explained that the institute "already exists here in deed if not in word."[106] He listed what he called the "haphazard" names under which this work had previously been carried out, such as the Rubber Research Laboratory, Office of Rubber Research, and Department of Rubber Research. One of the main problems was that the director of rubber research (Morton) had no autonomous department or organization. Rubber research took place within the existing structure of the wider department of chemistry. Morton said that the time had come to give this research activity "official life" and some autonomy from the chemistry department.

Morton pointed to the growth of rubber research activities in industry as an important indicator of future support. He argued that if this trend continued, it would require the establishment of a stronger institute within the university. Goodyear had already initiated a research program at the Rubber Research Laboratory, and Morton believed that "more industrial concerns will see the wisdom of having some of their fundamental, long-range problems studied at the university."[107] Morton also reminded Auburn and other key decision-makers that, in order to keep pace with the Soviet Union, an institute such as this would help to ensure progress in basic research and technical training.

Auburn strongly supported Morton's plans. In March 1956, however, he suggested changing the name of the proposed organization from the "Rubber Research Institute," to the "Institute of Rubber Research." Auburn preferred his title to Morton's because, he said, "[it] sounds a little more imposing to me."[108] Later that month, Auburn presented the plan to the board of directors. It named four main goals for the institute. The first was to advance the knowledge of natural and manmade rubber and plastics through theoretical research in polymer chemistry. The second was to continue to cultivate basic research contracts from the government, the military, and industry. Third was to assist the department of chemistry in teaching students with an interest in polymer chemistry. Finally, the new institute would work with the chemistry department to help train those who sought careers as polymer chemists or educators.[109] With the institute in place, Auburn believed that the University of Akron would become "the logical choice in any question involving basic rubber."[110]

Morton went on to describe the new institute's specifics. It would be staffed by a director, assistant to the director, and research chemists, associates, and assistants. Significantly, the director would no longer report to Thomas Sumner, the head of the chemistry department. Under the terms of the new institute, the director, who would be Maurice Morton, would report directly to the president of the university, Norman P. Auburn. This was the first step in establishing an autonomous polymer program at the University of Akron. An important relationship would still be maintained with the department of chemistry. Qualified staff members who worked at the institute could teach in chemistry, if their services were needed. Students in the traditional chemistry program would still have their coursework planned by the department. However, those students who sought advanced degrees in polymer chemistry would have their coursework planned through the cooperative efforts of the chemistry department and the Institute of Rubber Research.[111]

Financing for the new institute would follow the same model as the Rubber Research Laboratory. The laboratory had been essentially self-sufficient, with yearly grants totaling $150,000; they would be transferred to the institute. Any remaining funds would be reinvested into the operating budget and treated as contributions from the university to further growing the institute. This was important because, as Morton pointed out, should the board of directors approve this plan, it would

The Institute of Rubber Research. Maurice Morton (standing left), seated, left to right, are Irja Piirma, Howard Stephens, and H. James Harwood. Standing in back is Alan Gent. *Tel-Buch* (1963). *University of Akron Archives.*

require neither a subsidy from the university nor any tuition income to support itself. The university, which was already strapped for operating funds, could benefit tremendously from such an arrangement, as the international reputation of the institute would serve to increase the school's status both locally with the rubber industry, and internationally. The institute required no new facilities: since the Knight Chemical Laboratory was only five years old, Morton believed no additional space would be needed. There was room for growth in the Knight building, and additional equipment necessary for future contracts could be adapted to the existing facilities. This did not rule out eventual expansion. Should that need arise, Morton specified, the president of the university would determine the feasibility of a new building.

The board of directors was receptive to Morton's plan and on March 22 approved the Institute of Rubber Research.[112] One week later, Morton prepared a brochure to publicize the new organization. Auburn reviewed and revised the content, and in May Morton began sending copies to over three hundred companies throughout the rubber industry.[113] Morton sent personal letters with the brochures emphasizing that the two main objectives of the institute were to pursue basic scientific research and provide a postgraduate training ground for new specialists in polymer chemistry.[114]

Morton's message was that the new institute was carrying on the long tradition of rubber research at the university that began with Charles Knight. The brochure also highlighted the work of students and faculty in the laboratory with images of researchers determining the molecular weight of a polymer by osmotic pressure, measuring the particle size of an experimental rubber latex, discovering the stress-strain properties of experimental elastomers, and compounding an experimental sample on a rubber mill. Morton wrote, the "rise of synthetic rubbers has brought the whole science and technology of rubber into the broader field of polymer science."[115] An important change was taking place in an emerging discipline. A few weeks after the institute was approved, the U.S. Government Synthetic Rubber Research Program was terminated. As one era ended, another began with the institute and a new tradition in polymer science.

One important goal of the new institute was to improve the technical apparatus available for research in the late 1950s. Many believed the Knight Chemical Laboratory was in dire need of more advanced research tools. The Akron Rubber Group made several donations that resulted in new equipment. These included an electrically heated laboratory press that cured rubber and plastic samples and an Instron tensile tester. The Shell Chemical Company provided funds for the purchase of an internal mixer.[116] But the most important addition to the laboratory came in 1957 with Goodyear's gift of an electron microscope.[117] Scientists could actually see molecules, magnified to one hundred thousand times their original size by the microscope. Auburn and R. P. Dinsmore, vice president of research at Goodyear, made the presentation of the new equipment. "No one ever saw a molecule until the electron microscope was perfected," Dinsmore explained.[118] It enabled the direct viewing of large "super-molecules" of natural rubber and synthetic polymers. The microscope was put into immediate use, measuring the particle size of synthetic rubber latex. The journal *Rubber Age* reported that the "Institute is unique in possessing the necessary rubber processing and testing equipment required for the evaluation of the physical behavior of elastomers."[119]

The institute continued to attract industry and government contracts to the university. In 1958, Morton wrote that "an encouraging sign was the increase in participation by private industry."[120] He was most pleased with two new companies, Shell Chemical and United States Rubber, which had funded projects at the institute. The Shell contract, for $11,500, explored the mastication of synthetic rubber, while the U.S. Rubber contract, for $5,000, examined anionic polymerization. Auburn believed that the continuation of defense department contracts would also help to ensure the growth of the institute. In 1958, the institute was awarded a $78,500 contract (at the time its largest contract ever) by the U.S. Army Signal Corps to study the causes of the deterioration of rubber.[121] Despite this success, Morton said there was still a great need for increased funds, which were necessary to accomplish Morton's second goal for the future. In 1956, the school inaugurated, at virtually the same time as the Institute of Rubber Research, a new Ph.D. program in polymer chemistry. Finally, the dream envisioned by former president Hezzleton Simmons in the 1930s would be realized.

A PH.D. PROGRAM AND THE "FAMOUS FIVE"

Prior to the formation of the Institute of Rubber Research, the University of Akron employed a variety of rubber professionals within the chemistry department. Since the 1910s, these included professors, postdoctoral researchers, and master's students. But there had always been one major weakness. Graduate students who wanted to pursue the highest degree in chemistry had to leave Akron for universities that offered the doctoral degree. Neither the chemistry department nor the University of Akron as a whole was able to offer a Ph.D. on any level. Maurice Morton and Norman P. Auburn set out to change this situation by planning for a new degree offering—a Ph.D. in polymer chemistry.

In 1952, when Morton took over from Whitby and became the university's first professor of polymer chemistry, the Ph.D. became his central goal. He said, "The challenge for me then was this. . . . I felt that there was such a demand for higher education in polymers, we had to go beyond the master's degree. I set my sights on having a Ph.D. program. This was quite a challenge."[122] Morton believed this to be the first Ph.D. specifically in polymer chemistry offered anywhere in the world. Although Herman Mark's program at Brooklyn Polytechnic was granting Ph.D.'s in physics and chemistry with a concentration in polymers, Morton said, "We were the first to organize a Ph.D. polymer chemistry program."[123] By the time Morton retired, the city of Akron boasted the second highest proportion of all Ph.D.'s, per capita, to its population in the United States.[124] The rubber industry was drawing the talent to Akron, but for the first time, in 1959, Akron could produce its own.

The establishment of the Ph.D. program and the formation of the Institute of Rubber Research went hand in hand. The role of the institute was to provide the research laboratories and the funding necessary for a Ph.D. program. The scientists became mentors, teachers, and colleagues of the graduate students. The students used their work

One of the first polymer science-chemistry classes taught by Dr. Berry (1955). *College of Polymer Science and Polymer Engineering Archives. Photo by Tobias.*

as the basis for their dissertations and their own original research. Outside funding for the institute and research contracts provided the financial backing to make the Ph.D. program in polymer chemistry possible.[125] Morton believed that the combination of the institute and the Ph.D. program "can make the University of Akron the unchallenged leader in rubber science in this country, and, possibly, in the world." [126]

Although Auburn favored the Ph.D. program, he nevertheless wanted to be cautious. Since he had come to the university in 1951, he had advocated an "expansion in graduate work" as one of his primary goals for educational advancement. But in 1953, he said he had no desire to "upset the academic applecart." He knew the university's academic standards were high, and he did not want to detract from or "interfere with the nourishment of our great tradition."[127] Still, he feared complacency and wanted to plan for a dynamic future. Furthermore, he said that Morton was "not easy to dismiss."[128]

The polymer chemistry Ph.D. program had the potential to upset the established order of things at the school, but Auburn could neither resist nor deny the tremendous benefits such a program could bring. Nor could he resist Morton's enthusiasm for it. Auburn recalled, "He convinced us—me, the president's cabinet, the board of directors and the faculty—that we could become a leader in polymer science and engineering only if we could offer the Ph.D. degree."[129] To ensure that this was a sound academic plan, Auburn had Morton assemble an outside panel of polymer chemists, including Paul Flory, to offer their opinions about a potential Ph.D. program in Akron. They were unanimously optimistic about the opportunities it presented for the school, the rubber industry, and the further advancement of polymer science.[130] Flory would later help design the Ph.D. curriculum. Auburn gave his complete support to the planned program in 1956.

Auburn called the polymer chemistry Ph.D. not only an "educational milestone" but possibly the "most important academic advancement in the eighty-four-year history of Buchtel College and the University of Akron."[131] Ph.D.'s were rare in any discipline in the 1950s. In Ohio, only four other institutions offered them in any field—Ohio State University, Western Reserve University, Case Institute of Technology, and the University of Cincinnati. The University of Akron became the fifth.

Along with Auburn, the university council, the board of directors, and the North Central Association of Colleges and Secondary Schools accepted the decision. However, according to Auburn, the North Central Association's "first reaction was consternation [because] starting doctoral work in one specialized area was almost unheard of."[132] Several important factors led to the program's approval. First, the chemistry department faculty were uniquely qualified to lead students in their advanced research and studies. Second, the newly formed Institute of Rubber Research would provide essential funding, institutional support, and creative research opportunities for the students. Third, the scientific collection at Bierce Library and the library's association with the Rubber Division of the American Chemical Society offered students important nonlaboratory research data. Fourth, the newly built Knight Chemical Laboratory had the equipment and facilities necessary for practical experimental studies. Finally, the school's location in Akron and its proximity to the rubber industry, and Akron's reputation as the Rubber Capital of the World, would ensure an adequate supply of qualified students eager to enter the program. It was clear to all those who reviewed the proposed Ph.D. program that "no other institution of higher learning possesses such qualifications to offer graduate work in the restricted field of polymer chemistry."[133] The program officially opened its doors in September 1956.

The reaction to the program from corporations, the press, and university alumni was very positive. The rubber industry supported the program through several gifts, most notably those of William S. Richardson, president of B.F. Goodrich, who gave $30,000.[134] The press also responded enthusiastically. One journalist, in an article headlined "Ph.D.—Akron," wrote that the university was making an important "forward stride" with its new doctoral program and that its graduates would immediately earn the "highest respect" in the field.[135] Alumni reaction was also favorable. George Wilson wrote a letter to Auburn stating that "it is this kind of progress at the University that causes those of us who were graduated there to expand our chests a little further."[136] The chemistry department was also aware that the Ph.D. program would cause significant change. Thomas Sumner was more reserved in his reaction. "It is difficult to predict the impact that this venture will have on the Department," wrote Sumner, "and indeed, on the University, in future years, but it would seem that this impact should be profound."[137] Despite the positive feedback, the board of directors said the most important reaction came from the students. "Will we have any takers?" they asked.[138] Morton assured the board that there would be thirteen students beginning the program in September and that the goal was to graduate the first four in 1958.

The student response was positive, but it took a year longer for the first class to graduate than Morton predicted. In June 1959, the first five students (Ralph Milkovich, Edgar E. Bostick, Marvin A. Deisz, William E. Gibbs, and Masao Ohta) received their Ph.D.'s in polymer chemistry.[139] Watching these graduates receive their diplomas at commencement was one special guest. Charles J. Jahant had taken Charles Knight's first course in rubber chemistry and was a member of the class of 1909. Like many of Knight's students, he went on to a significant career in industry, becoming vice president and member of

General Tire and Rubber's board of directors.[140] Bringing Jahant, the traditional specialist in rubber chemistry, together with the five new Ph.D.'s in polymer chemistry symbolized the strong past and the start of a new era at the university.

In April 1960, after these first students graduated, the North Central Association of Colleges and Secondary Schools accorded the University of Akron full accreditation for its Ph.D. program. Its report stated: "The Ph.D. program in [polymer] Chemistry is strong. Financial support for it is adequate; the training of staff members is high; the space, facilities, and library holdings are satisfactory; and instructional provisions as evidenced by doctoral and master's theses, and qualifying examination questions are commendable."[141]

Maurice Morton began calling these first graduates the "famous five." Although he praised them all, he said that Ralph Milkovich "played a special role, indeed an historic role, at the University during those years."[142] Born in 1929 in Clairton, Pennsylvania, Milkovich earned his undergraduate degree in chemistry from Duquesne University in 1951. He spent the next four years working as a chemist at Koppers. In 1955, he returned to graduate school and began a master's degree in polymer chemistry at the State University of New York under the direction of Michael Szwarc. He quickly made a reputation for himself as a brilliant researcher. In 1956, he discovered "living" polymers, and his work appeared in a landmark paper in the *Journal of the American Chemical Society*. It demonstrated that polymerization of styrene by sodium naphthalene was actually a "living" polymer—polymer chains with the ability to propagate for a long time and grow to a specific maximum size while the degree of termination or chain transfer is negligible.[143] Milkovich both designed and constructed the all-glass high-vacuum apparatus that proved their existence.

Ironically, his success in making pioneering polymer discoveries did not carry as much weight as his Ph.D. qualifying examination. Though Milkovich was making significant contributions to the field in 1956, he failed to perform as well as his State University advisors had hoped on his examinations. Although Szwarc granted him a master's degree, he contacted Morton to see if he would be interested in Milkovich as a doctoral student. Morton enthusiastically accepted him. Morton argued

Ralph Milkovich is shown here doing polymerization studies using a high vacuum rack (c. 1960). *College of Polymer Science and Polymer Engineering Archives.*

that basic research ability in the laboratory far outweighed any student's difficulty in taking standardized tests. Milkovich arrived at the University of Akron at the most opportune time. He helped set up high-vacuum racks so that the other new Ph.D. students could begin work on anionic polymerization. In 1959, he completed his thesis, "Monodisperse Difunctional Polymers," and became a member of the first "famous five" to earn a Ph.D. from the University of Akron. He expressed his gratitude to Morton in his dissertation, thanking him for providing the "opportunity to study under his inspiring guidance."[144]

After graduation, Shell Chemical hired him to work in Torrence, California; he stayed from 1959 to 1963. There he became known as the "Johnny Appleseed of anionic polymerization." His living polymers were essential to such developments as thermoplastic elastomers, whose unique physical properties gave them the processability of plastics and the flexibility and durability of rubber. Milkovich helped to develop a triblock polymer (styrene-butadiene-styrene or SBS). When the polymer was heated above the glass-transition temperature of the polystyrene domains, it could be fabricated as a thermoplastic material. Once it cooled to ambient temperatures, it recovered its elastomeric properties. This new polymer was originally manufactured under the trade name Kraton, and Milkovich's name was first on the patent.[145] Its significance can be seen in the numerous plastic items in our daily lives, and thermoplastic elastomers are today a $2.2 billion industry.

After Shell, Milkovich transferred to General Tire and Rubber in Akron, where he worked from 1963 to 1969. During this time, his research focused on block copolymers and polymer alloys. Milkovich then moved to CPC International, in Argo, Illinois, where he helped to synthesize macromonomers, which became known by the trade name Macromers. He left CPC in 1974 when he joined ARCO Chemical. His primary focus there was the commercialization of Macromers, which he was partially successful in accomplishing. In 1984, he became the manager of new ventures for PPG Industries in Allison Park, Pennsylvania. He directed the research of new polymer programs until his untimely death in 1985. In 1988, the University of Akron held its Nineteenth Annual Polymer Conference, which also served as the Ralph Milkovich Memorial Symposium.[146]

Like Milkovich, Edgar E. Bostick, another of the "famous five," had a prolific career in the rubber industry. Before coming to Akron, he earned an undergraduate degree in chemical engineering from Auburn University. He worked at Goodyear Tire and Rubber until he volunteered to serve in the Korean War. He was a combat engineer with the X Corps, and the army awarded him a Bronze Star and the citation for Service Above and Beyond the Call of Duty. He returned to Goodyear in 1953 as a chemical engineer in the product development division. In 1956, he left industry to enter the Ph.D. program in its inaugural year. While at Akron, he received the American Society for Precision Engineering's (ASPE) Thomas T. Orr Award for meritorious work in plastics technology. He wrote his thesis on "Anionic Polymerization of Cyclic Siloxanes," and it won the *Rubber Age* Award for outstanding doctoral thesis in rubber chemistry.[147] After graduation, he remained at the university for one year as a postdoctoral fellow and then joined General Electric in Schenectady, New York. He worked there full-time for the next thirty years and as a consultant for ten years after that. By the end of his career, he had helped develop twenty-two patents for polycarbonate and siloxanes and contributed to another four hundred patents.[148]

The remaining three graduates also went on into careers in the rubber industry. Both Marvin Deisz and William E. Gibbs were born in Akron and received their bachelor's and master's degrees at the University of Akron. Like Bostick, Deisz was also a Thomas T. Orr Award winner, earning his recognition in 1952. Deisz's thesis was on the "Kinetics of Formation of Urethane Polymers," and while working toward his degree he taught the Experimental Physical Chemistry of Polymers and Physical Chemistry of High Polymers courses.[149] After completing his Ph.D., he went on to work at Shell Chemical in Emeryville, California. Gibbs's thesis was on "The Emulsion Polymerization of 2,3-Dimethylbutadiene-1,3." After graduation, he became a lieutenant in the air force. He became chief of the polymer branch of the materials laboratory at Wright-Patterson Air Force Base in Dayton. Later, as a civilian, he became an executive at Foster Grant. In 1987, the American Chemical Society presented him with its Distinguished Service Award.

Masao Ohta, the final graduate, was a native of Kobe, Japan, and earned a bachelor's degree at the Kyoto Imperial University (1942). For the next ten years, he worked as a research chemist in Tokyo at the Mitsui Chemical Company. He left for the United States in 1953, earning his master's degree from the University of California three years later. After his Ph.D., he went on to work at the Monsanto Chemical Company in Dayton, Ohio. These first graduates of the program served as important ambassadors for the university and helped both to inaugurate and spread the good news about the new polymer educational tradition in Akron.[150]

CONCLUSION—A NOBLE TRADITION

The significance of 1956 as a milestone year for the University of Akron could be measured in a number of ways. First, the university had championed one important response to the Cold War—advanced training in science for graduate students. As Auburn wrote, "America's well being, at home and abroad, in peace or in war, now and in the future, depends upon the quality, strength and vitality of its educational system."[151] The Institute of Rubber Research and the doctoral program in polymer chemistry exemplified this and would contribute in important ways, according to Auburn, to the well-being of the United States. A second measurement of the impact of the milestone year was through the quality of the Ph.D. graduates themselves. Here the university made immediate gains under Morton's leadership, the achievements of the "famous five," and those who came after them.

Another key way to measure the success of the Morton institutional and educational reforms was through their reception by established polymer scientists. Paul Flory, the 1974 Nobel Laureate in Chemistry, was one of these pioneers. He had extremely close ties with the University of Akron, even though he never taught or took classes there. Flory became friends with Morton, and as a colleague helped to initiate the Akron Polymer Lecture Group hosted by the university and design the Ph.D. curriculum. Even though Flory taught for nearly twenty-five years at Stanford University, he bequeathed his personal library to the University of Akron. It now resides in the Paul Flory Reading Room in the Goodyear Polymer Center.

Flory was not the only one of polymer science's "founding fathers" who enjoyed a relationship with the University of Akron. The most famous name in the field was Herman Mark. One of the key originators of the field, he gave the discipline of "polymer science" its name, founded the Polymer Research Institute at Brooklyn Polytechnic, and established the *Journal of Polymer Science*. In 1964, Mark wrote a letter to Maurice Morton congratulating him on Akron's "excellent new program on a Ph.D. in polymer science." Mark asked Morton to send him thirty copies of his program brochure so he could distribute them to his friends in the United States and Europe. Mark concluded by saying, "I wish I could come to Akron for two years and take all your courses—I could benefit from them."[152] (Ironically, Norman Auburn would visit Brooklyn Polytechnic, actually serving as acting president there in 1973.) Morton was very pleased with the letter. He quickly sent the brochures and complimented Mark on his contributions to the discipline of polymer science, writing "Shall we say that we are simply carrying on in the noble traditions that you yourself set for us a quarter of a century ago in this country?"[153] The start of Morton's own noble tradition had been a success. The 1960s and 1970s would be a time for him to cultivate and grow the seeds of the program he planted in 1956 and create a world-renowned polymer science institution.

4

RUBBER'S PAST AND POLYMER'S FUTURE

1958–1978

IN 1962, *TIME* MAGAZINE PUBLISHED AN ARTICLE IN WHICH the author recalled a scientific "age of innocence" when the pursuit of science was neatly divided into distinct specialties of chemistry, physics, and biology. Those days were gone forever, replaced by a new scientific reality of arcane and sometimes undecipherable subspecialties. One of the most important of these was the emerging discipline of "polymer chemistry." Although the *Time* author suggested that to most people this discipline meant "little or nothing," in the coming years polymer chemistry would play an essential role in everyday life and might even be the subject of "cocktail party chitchat."[1]

Maurice Morton was delighted by the piece, and immediately after reading it, he wrote a letter to President Norman Auburn, informing him that *Time* magazine had just proclaimed that polymer chemistry was "one of the new 'glamour' fields."[2] He made a special point of mentioning that it was the only field of chemistry the article discussed. Morton was entering an exciting phase of his career, and the growing national and popular exposure of the science he cared so deeply about reinforced his limitless passion and exuberance.

Throughout the 1960s and 1970s, Morton and Auburn continued to work together to innovate and improve the polymer program at the University of Akron. During these decades, a new interdisciplinary Ph.D. in polymer science was established (1964) and an autonomous Department of Polymer Science was created (1967). At the same time, Morton recruited a core of new faculty members who served as the foundation of a growing and increasingly interdisciplinary department. An expansion of research was made possible with the construction and opening of the massive Auburn Science and Engineering Center in 1968. During these years, the polymer program also strengthened its relationship with the Rubber Division of the American Chemical Society, welcoming it to campus, expanding its library, and working with the society to create unique educational opportunities.

However, Morton's boundless enthusiasm was not limited to polymer science. Several historical milestones were marked during the second half of Morton's tenure at the University of Akron, including the celebration of the fiftieth anniversary of the teaching of rubber chemistry, the annual inductions to the International Rubber Science Hall of Fame, and the twenty-fifth anniversary of the United States synthetic rubber program. Morton and the university both honored the past and planned for the future; their endeavors served to establish more completely the still emerging discipline of polymer science and the University of Akron's reputation as a pioneer in this increasingly "glamorous" field.

REMEMBERING THE PAST

On November 23, 1963, the world's rubber leaders had gathered in Akron, Ohio, to pay tribute to one of their discipline's founders, Wallace

Carothers. That day's symposium would honor the induction of the tenth member of the International Rubber Science Hall of Fame. But this anticipated event was overshadowed by a national tragedy. Exactly twenty-four hours earlier, an assassin's bullets had struck and killed President John F. Kennedy in Dallas.[3] Maurice Morton quietly took the stage of the Hilltop Room in the Gardner Student Center and addressed the somber audience, asking for one minute of silence for the fallen president. After the moment of reflection, Morton confessed that the previous day's "crushing news" left him with little enthusiasm for the induction. He had decided to proceed out of respect for the man whom they were about to honor, and those who had already traveled to Akron for the occasion. Morton found a parallel between Kennedy and Carothers—both had led brilliant and tragic lives. It is a "poignant reminder," he said, "of the irreparable loss we suffer when men of ability and leadership, in whatever walk of life, are cut down in their prime, be it by an assassin's bullet, or by other causes."[4]

The International Rubber Science Hall of Fame had been established in 1958 as a memorial to the pioneers of the growing rubber and polymer discipline. The idea emerged in a meeting between Norman Auburn and Ben Maidenburg, the editor of the *Akron Beacon Journal*. Maidenburg suggested that polymer pioneers deserved recognition, and he thought this could be achieved by establishing a hall of fame. Auburn was enamored with the idea and approached Morton and Whitby about it.[5] They favored the proposal, and the three quickly established a set of criteria for entrance into the new rubber hall. New members were inducted at an annual symposium that reflected upon the honorees' contributions to the field. No one still living was eligible for election, which not only prevented active political posturing, but also gave the occasion a more reverent air as those who had come before were honored.

The planning and organization of the hall of fame was a joint effort between the University of Akron and the Rubber Division of the American Chemical Society. Morton chaired the selection committee and shaped its membership. While some wanted representatives from the larger rubber companies, Morton thought that "direct participation by industry representatives may enlarge this committee unduly."[6]

Maurice Morton helped establish the International Rubber Science Hall of Fame. It was cosponsored by the University of Akron and the Rubber Division of the American Chemical Society. The Hall of Fame was located in the chemistry library in Knight Hall. Here Morton is pictured (center) with Thomas Sumner (far left) and G. Stafford Whitby (second from right). *Tel-Buch* (1959). *University of Akron Archives.*

He advised the committee to "think along international lines."[7] The first selection committee consisted of Morton, G. Stafford Whitby, and Thomas Sumner representing the University of Akron, and Ralph E. Wolfe and Arthur F. Juve representing the Rubber Division.[8]

The nomination committee recognized an international group. The first inductees were Henri Bouasse (1866–1953), Charles Goodyear (1800–1860), Ivan I. Ostromislensky (1880–1939), C. Greville Williams (1829–1910), and Carl O. Weber (1860–1905). Bouasse was a French physicist whose research on rubber elasticity in 1903 and 1904 paved the way for modern rubber physics. Goodyear, an American inventor, discovered sulfur vulcanization in 1839, which enabled a commercial rubber industry to emerge. Ostromislensky was a Russian chemist who was an early pioneer in studying polymerization and synthetic rubber. Weber, a German chemist, was the first to perform scientific investigations on vulcanization and the chemistry of rubber, while Williams was a British chemist who analyzed rubber in the 1860s and discovered the new compound "isoprene."[9] The portraits of the first inductees were hung in the chemistry library at the Knight Chemical Laboratory.

The first inductions to the International Rubber Science Hall of Fame coincided with another event that commemorated the past. In October 1958, the University of Akron planned an elaborate celebration to mark the fiftieth anniversary of Charles Knight's first course in rubber chemistry. President Norman P. Auburn chose the date because, he said, "According to all the records I have consulted, Dr. Charles Knight started the world's first program of study in Rubber Chemistry in 1908."[10] There was some confusion over the year. Charles Knight had been teaching some rubber chemistry to his classes throughout the early 1900s, but it was not until after the original Knight Chemical Laboratory was completed in June 1909 that he devoted an entire course to the subject. Although the celebration was a year premature, it offered for the university an opportunity to reflect on its history as the founding academic institution in rubber chemistry. Events included an all-campus convocation, the awarding of two honorary doctorates, a technical symposium, and a banquet. Most importantly, however, was the new International Rubber Science Hall of Fame's in Knight Chemical Laboratory.[11]

For the technical symposium, according to Morton, many possible themes were discussed, but "the final choice of subject was influenced by the historical development of rubber science."[12] While natural rubber had been important for over one hundred years, only recently had "giant strides in rubber science" taken place, with the rapid development in understanding polymers and the structure and behavior of macromolecules. Morton wanted to contrast these recent developments with the state of the field a half century earlier. At the symposium, three speakers discussed the past, present, and future of polymers.

G. Stafford Whitby offered a retrospective talk on rubber science. Peter Debye, a Nobel Laureate and professor emeritus at Cornell University, had begun work in the polymer field only recently, but had made important contributions, including the measurements of macromolecules. Paul Flory, then the executive director of research at the Mellon Institute, had researched for over twenty years the basic scientific principles that governed how macromolecules formed and behaved; he spoke on "Rubber Elasticity."[13] Maurice Morton presided over the symposium.

More than 450 guests attended the banquet that night; Alan Waterman, the distinguished speaker, discussed the importance of "research." Unlike many who espoused the space-race rhetoric to beat the Soviets in the Cold War, Waterman, in contrast, said, "We ought not to be too worried about matching them." He believed that vibrant academic research programs ensured that no other nation could attain anything more than a temporary advantage in the pursuit of science. The University of Akron, Waterman said, exemplified this belief. "Here we have a situation which typifies at its very best the partnership between the scholarly research of the universities and the growth of a major American industry. This partnership has made a notable mark upon the

As part of the fiftieth anniversary celebrations, Dr. Alan T. Waterman (director of the National Science Foundation) and Dr. G. Stafford Whitby (professor emeritus of rubber chemistry) were awarded honorary doctor of science degrees by President Auburn. From left to right, sitting in the front row, are Waterman, Whitby, Morton, and Thomas Sumner. *Tel-Buch* (1959). *University of Akron Archives.*

technological progress of the last half century."[14] With the festivities completed, Auburn congratulated those who had worked so diligently. To Maurice Morton, he simply said, "It did turn out well, didn't it?"[15] He also suggested that the Institute of Rubber Research would profit from the increased visibility.

With the fiftieth anniversary celebration in 1958, and the graduations of the first polymer chemistry Ph.D. students in 1959, Maurice Morton was ready to take on new challenges in the coming decade. His plans included assembling an eminent polymer faculty, establishing an Institute of Polymer Science, and more closely aligning the University of Akron with the Rubber Division of the American Chemical Society.

BUILDING A DEPARTMENT

In December 1962, Maurice Morton appeared before the University of Akron's Board of Directors. While his intent was to update the board on the Institute of Rubber Research and his plans for its future growth, Morton confessed he had an additional motivation. Top executives from four of the five largest rubber companies in the United States served on the university's board. "Lest you be alarmed, however, at the prospect of a sales talk or 'commercial,' " Morton assured the board that his message would be of vital importance for the future of the institute, the university, and the rubber industry.[16]

Morton explained that his primary focus was recruitment. In the early 1950s, Morton and Howard Stephens were the only full-time polymer chemistry faculty at the University of Akron. While the faculty grew marginally over the years, Morton's plans for a polymer institute could not be realized without expanding the department and bringing in a core of talented professors and researchers. "Our main need now and in the years to come," Morton told the board, "is the recruiting of an outstanding staff of polymer scientists in order to maintain and improve our standing as a center of polymer science."[17] This was important not only for the strength of the university, he explained, but because an academic center in polymers was necessary to keep Akron a destination for the brightest scientists and engineers in the country.

Many of those attracted by the polymer program would eventually serve the local rubber industry in Akron.

Morton believed that the "simplest and best" way to recruit outstanding polymer scientists was to ensure that they would have a high level of financial support.

Funding was essential to building a department, and Morton had demonstrated considerable success in this area. From 1956 to 1962, the Institute of Rubber Research had received $1 million in research grants. In comparison, from 1944 to 1955 grants totaled $500,000. Morton had effectively quadrupled available funding, and he believed this would be a central factor in assembling an elite polymer faculty in Akron.

This core faculty would become responsible for elevating the status of the department through their own research projects, as well as training the next generation of scientists. Morton hired nearly one key faculty member per year including H. James Harwood (1959), Irja Piirma (1960), Alan Gent (1961), Eberhard Meinecke (1962), Donald McIntyre (1966), Lewis J. Fetters (1967), and Joseph P. Kennedy (1970). Morton demonstrated an uncanny ability not only to attract quality professors to the university, but also to create an institution that could retain them by establishing a productive, supportive research environment. Many of his faculty remained at Akron for thirty years or longer.

One of H. James Harwood's first interactions with Morton was as an undergraduate student in 1951. Morton had entered his physical chemistry class and berated one of the students for reading a refractometer incorrectly. This device measures the amount of light that is bent, or "refracted," as it passes from air into a sample. Harwood said that at the time the experimental facilities were very meager, and the refractometer was one of the major instruments at the university. Therefore, the incorrect reading of it was of particular irritation to Morton, who was never one to hide his emotions.[18] Harwood received his undergraduate degree in chemistry at Akron, and then he earned a Ph.D. in organic chemistry at Yale in 1956. Three years later, Morton invited him to return to Akron to become a member of the Institute of Rubber Research. Harwood was one of the first polymer synthesis scientists to use a computer as an experimental tool. He

H. James Harwood. *Polymer* (Fall 1996).

was an enthusiastic user of the school's first computer, which was an IBM 1620 that had 16K of memory and used paper tape for data input. In the mid-1960s, he wrote FORTRAN programs to analyze light scattering and gas chromatography data and calculate the structural features of copolymers.[19] Harwood enjoyed a dynamic research and teaching career at the university, supervising the work of ninety-one Ph.D. students, holding thirty-one patents, and writing nearly two hundred research papers.

Morton's next hire, in 1960, was Irja Piirma, who, like Harwood, already had a relationship with the university. In 1952, Piirma became Maurice Morton's research assistant in the department of chemistry. Ten years earlier, she had earned her Diploma in Chemistry at the Technische Hochschule Darmstadt, in Germany. She continued her graduate studies in Akron during the 1950s. She earned her Ph.D. in 1960, and Morton asked her to remain on the faculty.[20] Three years later, she became an instructor in chemistry, where she also continued her own research in the kinetics and mechanism of emulsion polymerization. She became a professor of polymer science in 1981 and served as department head from 1982 to 1985, achieving international acclaim for her work on polymerization kinetics and polymer colloids.

Irja Piirma. *Polymer* (Winter 1993).

Morton's third major hire was Alan Gent, who joined the faculty in 1961. Though he started his career working in British rubber factories after World War II, research soon drew his attention. While earning his Ph.D. from the University of London in 1955, he

Alan Gent. *Polymer* (Fall 1992).

worked as a research physicist from 1949 to 1958 at what is now known as the Malaysian Rubber Producers' Research Association, eventually leading a team of seven scientists and technicians there and becoming its principle scientist. Morton, aware of Gent's growing expertise, thought a physicist would be a valuable asset to his program so he visited Gent in England in 1961 and offered him the position of professor of polymer physics.[21] Gent recalled that this was an "attractive position" and decided to "gamble on it."[22] At Akron, he thrived in the academic environment and immediately realized how he had previously been "confined by his employer." Gent concluded that, unless you were a professor, you could never really do research. He enjoyed the advantages of his academic freedom, studying the strength of adhesive bonds and the failure mechanisms in rubbery, glassy, and crystalline polymers. By the end of Morton's tenure, Gent had become the first winner of the Mobay Award (1964), the Bingham Medal of the Society of Rheology (1975), and the Colwyn Medal of the Plastics and Rubber Institute (1978). He also won the Charles Goodyear Medal in 1990 from the American Chemical Society's Rubber Division and was inducted as a member of the National Academy of Engineering.

Like Gent, who originally planned to stay at Akron for only two years, Eberhard Meinecke arrived in 1962, intending to stay only one year. Both men remained at the University of Akron for over thirty years.

Eberhard Meinecke. *Polymer* (Fall 1992).

Meinecke, a German, earned his doctoral degree in engineering at the Braunschweig Institute in 1950, in his native town. He came to the United States in 1950 on a NATO fellowship and stayed for a postdoctoral fellowship at the Polymer Research Institute at the University of Massachusetts. At Akron, Meinecke specialized in the applications of rheology, the mechanical behavior of plastics and rubber, and the effect of this behavior on the processing and performance of polymers. Soon after their arrivals, Gent and Meinecke began working together. In 1965, they received a grant from the National Science Foundation to examine the potential of substituting rubber for metal in mechanical spring systems. "One of the principle difficulties in using rubber springs," according to Gent, was "the lack of basic information about their elastic response and durability under various conditions."[23] The two men engaged in materials research to determine how various types of rubber would respond in strength and vibration tests. This was some of the earliest work in polymer engineering at the university. Since that time Meinecke published over one hundred papers on viscoelasticity and optical properties and processing. He believed in the importance of a strong academic/industrial relationship as a way to further his own research and also inform students. "If we were not consultants to companies, then two things would happen," he wrote. "It would be an indication to others that we do not know what we are teaching, and we would not gain valuable experience outside of the 'ivory tower' for teaching and research."[24]

Donald McIntyre. *Polymer* (Fall 1996).

Other key funding opportunities followed for the polymer faculty and their Ph.D. students. In 1965, the University of Akron attempted to become an official "center of excellence" in polymer science. This enabled the Institute of Rubber Research to apply for a National Science Foundation (NSF) grant of $2,133,000. The NSF had already established thirteen other "centers of excellence" in science at other colleges and universities across the United States. If Akron was accepted, it would become the first in polymer science. A six-person team visited the University of Akron on behalf of the NSF to evaluate the program's worthiness.[25] The university became a "center of excellence," and the resulting funding was a boon to its research program.

Morton had not finished adding new faculty to his institute. Donald McIntyre arrived in 1966. His initial interest in chemistry grew out of his experiences in high school in Detroit. During the Depression, several Ph.D.s from local industries lost their jobs in industry and resorted to teaching in high school. These scholars inspired McIntyre to pursue a career in academic research. He earned a doctorate in physical chemistry from Cornell University in 1954 and then worked at the National Bureau of Standards (what is now the National Institute of Standards and Technology) for ten years, from 1956 to 1966. McIntyre was always impressed with Morton and learned a great deal from him. "Morton," he said, "knew how to sell himself and his ideas, and he was untiring in his efforts to build the program."[26] McIntyre's own research focused on the thermodynamics of macromolecules, the characterization of polymers, and natural rubber.

Lewis J. Fetters became a professor of polymer science in 1967. He earned his Ph.D. from the University of Akron in 1963, and after his postdoctoral fellowship and position as a chemist at the National Bureau of Standards, Morton convinced him to return to Akron as a professor of polymer science.[27] At Akron, Fetters researched the kinetics of anionic polymerization, block polymers, and the characterization of macromolecules by solution methods, viscosity, and light scattering.

Joseph P. Kennedy's road to the University of Akron was perilous. During his final undergraduate year in chemistry at the University of Budapest in 1948, the post-World War II communist tensions became unbearable.[28] Because Kennedy had been passing out pro-United States leaflets, he was called before a government review board, and although he was released, he knew he would have to leave Hungary. Two days later, he escaped into Austria, where he earned a Ph.D. in biochemistry at the University of Vienna. He continued his research at the postdoctoral level at the Sorbonne and McGill University, followed by eleven years in industry at Esso (which is now Exxon) in New Jersey. In 1970, he joined the faculty at Akron. By the early 1990s, he had published over five hundred scientific articles and reviews, and held fifty-six U.S. patents.

In the midst of his faculty recruiting, Morton also worked toward restructuring the Ph.D. program, refocusing it away from chemistry and toward an interdisciplinary approach. At the same time, Morton was also preparing a new administrative structure which included the transition from the Institute of Rubber Research to the Institute of Polymer Science and the creation of an autonomous Department of Polymer Science. Without the new faculty serving as a core of the University of Akron's polymer science program, none of these achievements would have been possible.

Joseph Kennedy. *Polymer* (Fall 1996).

RUBBER OR POLYMERS?

With the growing faculty, Morton began to realize his larger polymer visions. When the Institute of Rubber Research was established in 1956, it was clear that the rise in importance of synthetic polymers necessitated fundamental basic research. By the mid-1960s, the emphasis was on "polymer science." At the surface, Morton knew this required name changes for the Institute of Rubber Research and the Ph.D. program in polymer chemistry. But there were essential structural changes as well, including an institutional separation from the department of chemistry and a new interdisciplinary Ph.D. program.[29]

In May 1964, Morton assembled a group of research directors at local rubber companies to discuss plans for a new interdisciplinary program in polymer science, as well as a name change away from "rubber."[30] Morton called this the "question of 'rubber' vs. 'polymer,' " and in the end the consensus was to adopt the latter term.[31] Morton acknowledged that the term "rubber" had traditionally been associated with Akron, but he believed it was important to distance the program from it. He said, the "use of the word rubber imposes a 'severe restriction' on public understanding of the Institute's activities."[32] The rubber companies had themselves evolved into a new polymer industry concerned in large part with producing new plastics and fibers. The name of the Institute of Rubber Research became "The Institute of Polymer Science."

Academically, Morton was making changes as well. In 1964, the polymer chemistry Ph.D. program broadened to become an interdisciplinary program in polymer science, a change made due to the nature of polymers themselves. The ultimate aim of the polymer discipline is to understand the relationship between the macromolecular structure and physical behavior. Morton believed that to achieve this understanding students would have to "cut across the usual boundaries between different physical sciences."[33] Organic chemists might explore the synthesis of new long-chain polymers. Physicists might work toward characterizing the size and structure of the new molecules. Engineers were required to discover how to process and manipulate the materials in real-world applications. Morton wanted to bring these three polymer activities together academically. He restructured the polymer chemistry program into an interdisciplinary program in polymer science with three areas of specialization—chemistry, physics, and engineering.

In November 1964, Morton announced his interdisciplinary program to the academic and industrial community. He created new informational literature and sent copies to all the institute's sponsors. For example, to Paul H. Johnson, the director of research and development at Firestone, he wrote that he planned to "publicize this program widely" to colleges and universities throughout the United States and the world, with the goal of attracting the best and brightest graduate students in science and engineering to Akron. Morton emphasized the uniqueness of the program, writing "We believe that we are the first institution in the United States to inaugurate this type of polymer science program."[34]

By 1966, the University of Akron had several other polymer graduate programs competing against it. In the United States, these included the Case Institute of Technology, the University of Massachusetts, the Polytechnic Institute of Brooklyn, Rensselaer Polytechnic Institute, and the Stevens Institute of Technology. Of these, three were part of engineering colleges (Case, Rensselaer, and Stevens), while the Brooklyn program was aligned with the chemistry department. Only Massachusetts was attempting an interdisciplinary approach, and Morton speculated that it was "frankly patterned after our program."[35] Furthermore, though Massachusetts attempted to create an interdisciplinary program, in Morton's opinion it was successful in incorporating chemistry and physics into the program, but not engineering.

Morton was convinced the unique approach the University of Akron was taking was preferable. He said that the "ace-in-the-hole" was its interdisciplinary Ph.D. program, which was unparalleled because of the administrative obstacles and internecine rivalries that got in the way of bringing these different academic pursuits under one roof elsewhere. The University of Akron had overcome these hurdles and by 1967 had officially established its own Department of Polymer Science. But, despite the strengths and uniqueness of the polymer program in Akron, Morton warned about increased competition in the future. He said, "Having reached the 'big league' we must now 'think big.'"[36] One way of "thinking big" was for the university to terminate its status as a municipal school and adopt new state status.

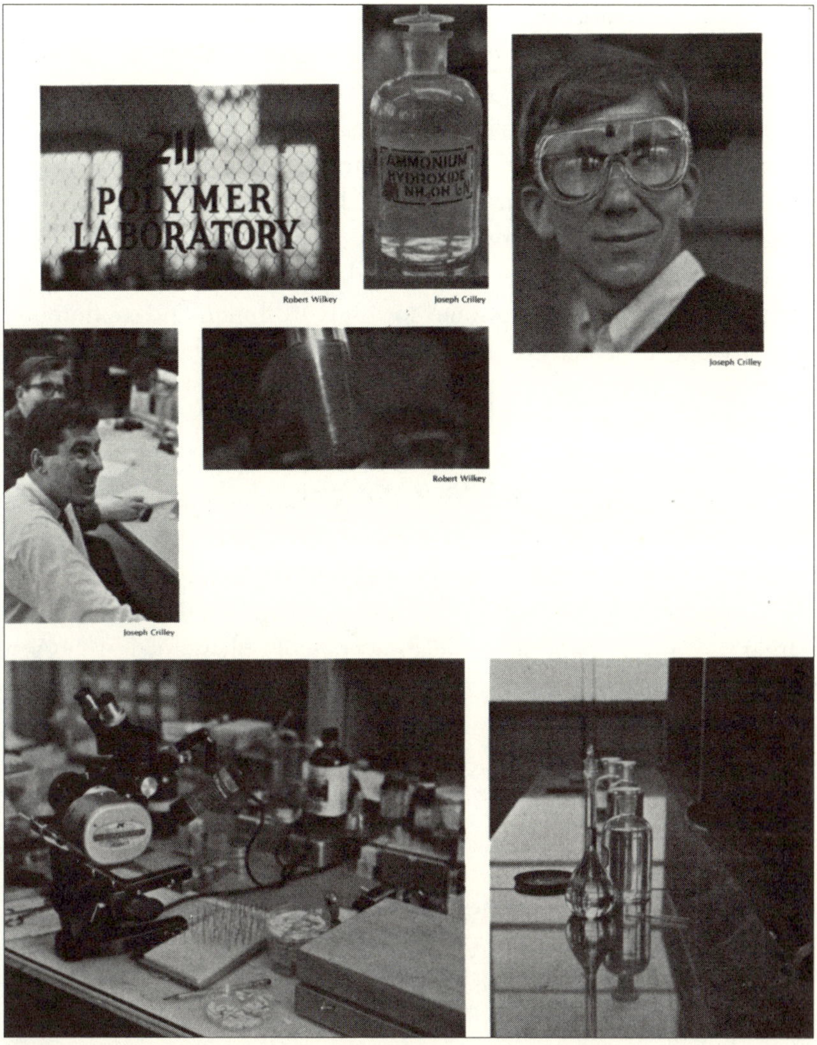

Polymer Laboratory at the University of Akron in 1967, the year Maurice Morton officially established the Department of Polymer Science. *Tel-Buch* (1967). *University of Akron Archives.*

A POLYMER GIANT

In the late 1950s, the University of Akron needed money. It was still a municipal school, and though enrollments were increasing, the city allocated little new tax money to support its growth. A larger student body meant that more buildings, instructors, equipment, and books were needed, and these needs were growing at a greater rate than tuition income. President Auburn, cooperating with municipal university leaders in Toledo and Cincinnati, made a case for state funding. But their efforts encountered little support from Governor Michael DiSalle. Their fortunes changed in 1962, when James A. Rhodes became the state's new governor. He immediately signed into law a bill giving the university some state subvention. The bill, signed on July 1, 1963, allocated a fixed amount of money for every freshman and sophomore, roughly 8 to 11 percent of the school's operating budget. While helpful, this did not solve the funding crisis, and school officials began pressing for a greater state commitment.[37]

In October 1963, Maurice Morton returned to Columbus to argue for further state aid. The Institute of Rubber Research was, in his words, "an important and unique 'natural resource' of the State of Ohio." Despite its growth and ability to secure industrial support, it still lacked the funds to equip laboratories with the experimental devices necessary if it was to remain on the leading edge of polymer research. As a result, because the institute lacked funds, it had to turn away qualified students and restrict recruiting for advanced faculty. Morton wrote, "It is timely, therefore, to consider direct aid from the State of Ohio to stabilize and expand the activities of the Institute."[38]

Norman Auburn also demanded greater state commitment and would not stop until he achieved his goal. But not everyone favored this change, and Auburn met with opposition. Existing state universities in Ohio did not welcome another "mouth" to feed from the available funds. State Democrats opposed the idea because they believed it would require significant new taxes on Ohioans. In Akron, some alumni resisted giving up the municipal status that had saved the school during its transition from Buchtel College in 1913. Despite the adversity, Rhodes and Auburn won the battle, and the Ohio General Assembly signed a bill on August 12, 1965, that began the process of converting the university to a state school. On December 8, 1965 the university's Board of Directors approved the decision and began planning for the transition. In May 1966, Akronites voted in favor of the new status by a margin of 36,034 to 4,015. Many believed this to be the most decisive vote ever in Summit County history.[39] Auburn would later be known as the "architect of state status."

State status resulted in numerous changes across the campus. This included a larger tax base with revenues that rose with enrollment and a more diverse student body. But, most significantly, state status guaranteed that the university could begin expanding its facilities. In 1963, a state bond issue had passed that allocated $6,000,000 for capital improvements. The Board of Directors decided these funds should be earmarked for a new science and engineering center. Rhodes and university officials believed this improvement would be an opportunity not only to increase the level of scientific instruction and research, but also to create new industry jobs.

In October 1965, ground was officially broken on the Auburn Science and Engineering Center. According to historian George Knepper, "The structure that emerged was a giant."[40] Some speculated that it was the largest academic building housed under one roof in the state. Including the parking decks below, the main building measured a total of five acres. It provided an immediate increase in laboratory, research, and classroom space, and supported over two hundred research labs. The College of Engineering soon moved in and used the facilities for its new Ph.D. program started in 1969. The building also enabled the expansion of graduate studies in biology and mathematics. And the Auburn Center became the new location of the Institute of Polymer Science. The institute occupied the north tower, with forty thousand square feet of space, along with Whitby Hall, which added another twenty-five thousand square feet.

President Auburn wanted the $10.5 million building to make a strong statement about the school. He reminded everyone involved in the dedication proceedings that this "may be one of the first important public functions of the University as a state university."[41] In November 1966, Maurice Morton became a key member of the faculty committee planning the dedication ceremonies for the Auburn Science and Engineering Center.[42] His goal was to "organize an attractive and prestigious program that will appeal to the whole 'polymer community' around the country."[43] He nominated W.O. Baker, the vice president for research at the Bell Telephone Company, to be the keynote speaker.

As one of the few polymer scientists who was a top executive, Baker had a deep appreciation of science and displayed a thorough understanding of human affairs.[44] In his recommendation letter, Morton said he was confident that Baker would present a dignified, scholarly, and witty address. These ceremonies were extremely important to Morton because, as he said, "I am particularly anxious to use this occasion, which represents a milestone in the growth of our Institute, to boost our image both nationally and locally."[45]

The university also wanted to use the occasion to mark an important historical milestone. The opening of the new Auburn Science and Engineering Center coincided with the twenty-fifth anniversary of the U.S. government copolymer research program, which had played such an important role during World War II.[46] Individuals like Waldo Semon from B.F. Goodrich helped to put together names of key individuals associated with the program in the 1940s.[47] Then Maurice Morton contacted members of this group to inform them of the celebration. From the very beginning, Morton received a "significant and enthusiastic response."[48] Whitby was pleased that the university was taking the time to remember its important contribution toward winning World War II. The dramatic growth of the synthetic rubber industry in the postwar period, Whitby said, was "a testimony to the soundness of the research and development program of which the twenty-fifth anniversary is now being celebrated."[49] Over two hundred participants from the synthetic rubber program attended the celebration.[50]

The dedication of the Auburn Science and Engineering Center took place on April 19, 1968, in a daylong ceremony. Ohio governor James A. Rhodes attended, returning for a third time after previous visits for the groundbreaking and cornerstone-laying. W.O. Baker presented the main address, as planned, at the convocation held in the Memorial Hall gymnasium.[51] Just as Morton intended, the ceremony brought together a diverse group of presenters from around the country.[52] Over five hundred people attended the ceremonies.[53]

The dedication of the building went well. Auburn called it a "splendid program."[54] W.O. Baker wrote that it succeeded in bringing "attention and interest to an estimable institution and to an important epoch in our national science and technology."[55] After the ceremony, Baker spoke with Harvey Firestone Jr. about the polymer program at the University of Akron. They agreed it was making a significant contribution to linking academia and industry. While state status promised a sound financial future for the university, the Auburn Center provided an opportunity to grow the polymer program in new ways, such as a partnership with the Rubber Division of the American Chemical Society. By the end of Morton's tenure, these two organizations would have officially joined forces in several important educational, administrative, and intellectual pursuits.

THE RUBBER DIVISION PARTNERSHIP

In 1874, the same year that the first college-level science course was taught in Akron, the chemistry profession was celebrating a milestone. It was the centennial of Joseph Priestley's discovery of oxygen, and a group of influential chemists used the occasion to discuss the formation of a new professional society. While the chemists all agreed that they needed their own society, there was a split between those who favored a distinct organization and those who preferred to become a chemistry subsection of the American Association for the Advancement of Science (AAAS). For two years, proponents of the chemistry subsection won out, but it became clear that the growing profession needed its own society. During an April 1876 gathering at the College of Pharmacy in New York, John W. Draper became the first president of the new autonomous chemical society. W.M. Habirshaw, a well-known chemist in the rubber industry, became its treasurer. One year later, the American Chemical Society officially incorporated in the state of New York.[56]

As the American Chemical Society grew, it became more specialized, and groups of chemists with similar professional interests formed their own sections. In 1909, Charles Goodrich, with twenty-eight other men, established the India Rubber Chemistry Section; Goodrich became the first chairman. The section's initial goal was to share information about rubber research and provide a common meeting area to discuss problems. It took several years to overcome the inherent secrecy within the industry, but by 1919 the section was making real progress. Rubber testing and analytical procedures were being standardized, and the experiences of World War I and the growth of the automobile industry were increasing the pace and necessity of rubber research.

Because of the success of the India Rubber Chemistry Section, the Council of the American Chemical Society began discussions to organize a new Division of Rubber Chemistry. There was an important distinction between a "section" and a "division." A section could not authorize any action on its own without the approval of the parent American Chemical Society. Furthermore, overall supervision of the section was also under the control of the parent body. Division status, on the other hand, would enable the rubber chemists to have their own permanent organization that elected its own officers, enrolled its own members, and authorized its own actions.[57] The first meeting of the Division of Rubber Chemistry was held in Philadelphia in October 1919.

The Rubber Division grew rapidly, and by the late 1920s its leaders advocated the formation of local chapters. In January 1928, the New York group held its first local meeting, followed by Akron (February 1928), Los Angeles (May 1928), Boston (November 1928), and Chicago (June 1929). Many believed that the establishment of local groups would not only enhance the technical capabilities and geographical spread of the Rubber Division, but it would also continue to foster the spirit of communication. Another important communication strategy was launched in 1928—the first issue of the division's journal, *Rubber Chemistry & Technology*.

From the start, the University of Akron played a role in forming and cultivating the Rubber Division. The primary technical concern of the rubber industry during those early years was the quality and measurement of rubber. Natural rubber came from South America, Central America, and Central Africa, and there were numerous varieties. In 1909, *India Rubber World* published a list of varieties that included thirteen types of Para rubber, twelve from Africa, eight from Central America, three from East India, and one from Ceylon.[58] It was of great importance for the rubber compounders to understand the material differences in rubber from different parts of the world. Akron professors played a seminal role in attacking this problem within the structure of the Rubber Division. Charles Knight chaired the committee on Testing and Rubber. Later, in 1919, Hezzleton Simmons gave an important report on his recommendations for tests on vulcanized rubber, and tensile and endurance tests.[59] Simmons also served as secretary of the Rubber Division for nine years, a position that represented a continuity of vision; a new chairman rotated in and out each year.

G. Stafford Whitby also served a unique role in the early Rubber Division. While his technical and scientific skills were always in demand, he also used the society to showcase some of his other talents. For example, in the April 1923 meeting at Yale University, Whitby was a member of the entertainment committee that performed a skit after dinner. He and his fellow players dressed as rubber industry pioneers, with Whitby impersonating Thomas Hancock. He tried to entice the audience in a sing-along to what he thought was a well-known British song, but he wound up performing a solo because no one else in attendance had ever heard it before.[60] While the University of Akron and the Rubber Division interacted only through the efforts of its professors during those early years, by the mid-twentieth century the affiliation became more official.

The partnership began with the Rubber Division library. There has always been a close link between libraries and laboratories because effective laboratory research depends upon books and periodicals for research.[61] Soon after the formation of the India Rubber Section in 1909, Hezzleton Simmons suggested that the rubber industry needed a library devoted to rubber research. He supported efforts to establish a "special library for the rubber industry" at the University of Akron—no easy task, but the need for the library grew every year. In 1920, Josephine A. Cushman, the associate librarian at the university, said the rubber industry suffered because its relevant research materials were "scattered throughout almost the whole field of chemical and physical publication."[62] The proposed rubber library at the University of Akron would bring these disparate resources together in one place for the rubber chemist. The first step was to purchase all rubber-related scientific abstracts, such as *Chemical Abstracts*, and then attempt to build a complete collection of the associated journals themselves. But collecting the materials was only half of the battle, and ongoing administration (cataloging and indexing) was considered the key to making the library useful for the scientists. Although the larger plan to establish a central library for the entire Akron-based rubber industry never materialized, the effort to concentrate more specifically on acquiring rubber journals and books was incorporated into the larger university library system.

After World War II, the growth of literature was expanding at such a rapid rate that the Rubber Division wanted to establish an official library and revisited the plan from the 1920s.[63] Instead of building a new facility, it selected the University of Akron as the library's location because it already had a large rubber collection. The original recommendation for this plan came from R.P. Dinsmore, chairman of the Rubber Division's finance committee, in September 1946.[64] At an executive committee meeting, he said, "It is recommended that consideration be given to the establishment of a Rubber Division Library to be located at and operated by The University of Akron." One of Dinsmore's primary goals was that the Rubber Division would allocate funds to microfilm documents related to the rubber industry and make them available to its members. Projection equipment for viewing the documents would also be purchased. The Rubber Manufacturers Association provided financial support for the library project.[65]

In March 1947, the Rubber Division's library committee held its first meeting. It agreed upon an ambitious mission—to "help The University of Akron build up the most complete rubber library in the world."[66] Exactly one year later, the Rubber Division reported that the local rubber companies were very willing to support the creation of such a library. Those consulted included representatives from Firestone, Goodrich, General Tire, Goodyear, and U.S. Rubber. By 1949, the library began part-time operation with a librarian whose responsibilities included investigating references, looking for articles on specific subjects, and developing special bibliographies. The demand for these services was so great that the Rubber Division expanded by hiring a new librarian, Betty Jo Clinebell, in 1952—the first year the library operated on a full-time basis.[67]

In 1965, the University of Akron established its Center for Information Services (CIS), a major development for the Rubber Division

Library. Panos Kokoropoulos served as the initial acting director of what was then the largest computer resource for rubber and plastics information in the world.[68] The database indexed 350 American and foreign scientific journals relating to synthetic rubber, natural rubber, and plastics. In 1976, the library changed its name to the John H. Gifford Memorial Library to honor a past chairman and staunch supporter of the library. Gifford had been chair in 1971 while continuing his industrial work with Continental Carbon. The Gifford Library is now located in the Auburn Science and Engineering Center, as part of its Science and Technology Library.

While the library was becoming an integral part of the university campus, another key association with the Rubber Division was being forged. In October 1962, Maurice Morton invited Gilbert H. Sewart (Rubber Division chairman elect) and Dale F. Behney (Rubber Division treasurer) to an after-hours meeting at the University Club. Sewart worked for General Tire and became chairman of the Rubber Division in 1961. Behney, of Harwick Standard Chemical, would serve as chairman of the Rubber Division in 1967. Morton, Sewart, and Behney met to discuss a plan to locate the Rubber Division's administrative offices permanently at the University of Akron. After the meeting, Morton was encouraged, and wrote in a private letter, "It appears that both of the above gentlemen are strongly in favor of this proposal, and that they feel that there should be sufficient backing for it at their forthcoming executive meeting."[69]

This issue was of increasing importance. For fifty-four years, the Rubber Division had no central location, with the frequent result that administrative duties would overtake one of the Rubber Division's officer's homes. For example, in the 1950s *Rubber Chemistry & Technology* had escalated in importance, and its editor, C.C. Davis, struggled to keep up with its growth. His home was taken over by the mountains of paperwork associated with the publication, including correspondence and submissions. As one report noted, "it was obvious that no slackening in the tide would occur."[70] Over time, administrative demands increased with more members and a broader array of activities at the division, so establishment of permanent offices became a more pressing concern. Rubber Division administrators looked at the University of Akron as a natural location for this office, and it opened its doors there in 1966.[71] At that time, only one person, Marge Bauer, staffed the office. Her equipment consisted of a desk, telephone, and typewriter. Despite these modest beginnings, Bauer remained at the university for over twenty years, and by the time she retired she directed a staff of five people in Whitby Hall.

Bauer believed one of her main challenges was accommodating the distinct needs of the different chairmen. She said, "Each new 'administration'—and there is a new chairman and many new committee heads each year—has a slightly different way of operating. We have to adjust to that and at the same time keep things moving smoothly."[72] Bauer and her staff were responsible for managing the subscriptions and letters to the editor for *Rubber Chemistry & Technology*, coordinating all membership information, collecting dues, updating the directory, tracking all communication relating to attendance and exhibits at Rubber Division meetings, and managing the finances for the entire Rubber Division.

Another important relationship between the University of Akron and the Rubber Division began in 1962 when the division started a correspondence course in rubber technology. For years, the Rubber Division had been involved in preparing and presenting courses to industry. These courses were so important because there were still few opportunities to learn more about rubber chemistry at American universities. The industry relied upon the division for basic training.[73]

The plans for a correspondence course began in 1961 when C.V. Lundberg, chairman of the Rubber Division education committee, was attending a Louisville meeting of the organization. There the Philadelphia, Chicago, and Southern Rubber groups expressed an interest in establishing a correspondence course in rubber chemistry. Many of the rubber groups had in the past offered their own courses on rubber technology, but it was a challenge to get many people to sign up because it was difficult for group members to travel to one centralized location for weekly lectures. A correspondence course (instructors would send assignments, and students would return homework via mail) could offer advanced training to the geographically separated members of the Rubber Division. The question was what academic institution could manage such a program? Lundberg immediately thought of Maurice Morton. "I am writing to you," Lundberg wrote Morton, "to see if the University of Akron, you personally or some other qualified person at the University might be interested in setting up and conducting such a course."[74] Lundberg and the Rubber Division thought the University of Akron would be a perfect partner in this endeavor for several reasons: the campus was in the heart of the rubber capital, the university had demonstrated expertise with professors like Maurice Morton, and the Institute of Rubber Research had a desire and ability to teach rubber chemistry to a wider audience.

From 1955 to 1962, Morton and the university had already instituted a number of innovative courses for the rubber industry. One course began in 1955 for tire dealers, the first educational program in the nation targeted to this group.[75] In 1956, Morton served as the director of another new annual course entitled the "Chemistry and Physics of Elastomers."[76] It covered topics such as new developments in Hevea latex, the chemistry of synthetic rubbers, the chemistry of vulcanization, and the physical behavior of elastomers. Physical behavior topics included elasticity theory, dynamic properties, relaxation, tear, abrasion, tensile strength, and reinforcement. In the early years, interest in the course was so great that dozens of applicants had to be turned away.[77]

Another unique program was a six-week series of special courses developed in collaboration between the University of Akron's Institute for Civic Education and the United Rubber Workers (URW). The series, entitled "A Chance to Learn," began in 1959. Instead of focusing on rubber research, it attempted to provide general educational skills to those working in the rubber industry. These included effective speaking, teaching, reading and writing skills, and social issues in recent American history. The course was also designed to make these workers better union members, and the curriculum included primers on labor legislation and parliamentary procedure. Although none of the polymer faculty was involved with the program, the initiative demonstrates another link between the university and the rubber industry.[78] L.S. Buckmaster, the URW International president, believed that a strong union required cultivating leadership skills through education. He hoped this partnership with the University of Akron would provide useful training.[79] By 1961, over two hundred union members were signing up for the courses on a yearly basis.[80]

In 1961, the university also offered a series of lectures on tire engineering sponsored by the university and the area rubber companies, including Firestone, General Tire, B.F. Goodrich, Goodyear, and Seiberling Rubber.[81] One year later, in 1962 the University of Akron, in conjunction with the Akron Rubber Group, inaugurated another series for the less technically oriented, an eleven-part weekly lecture on the rubber industry for sales, purchasing, and marketing executives. [82] The Polymer Lecture Group also continued to operate throughout these years. Its annual series of lectures brought academic and industrial leaders to Akron to talk about pioneering research.[83]

It was in part because of Morton's track record with these various courses—the 1955 tire dealer course, the 1956 chemistry and physics of elastomers course, the 1959 URW Chance to Learn series, the 1961 lectures on tire engineering, and the continuing educational offerings by the Polymer Lecture Group—that the Rubber Division approached him for the correspondence course. Morton was very interested in Lundberg's idea and wanted even more input into the design and teaching of the course. He wrote to Lundberg, "it might be feasible for myself, with the assistance of some of my staff, to actually set up a study program for this course."[84] Morton proposed using the existing library staff to assist with the clerical, correspondence, and other technical information, as he strongly favored centralizing all of the courses organization within the university. Lundberg was pleased with Morton's enthusiastic response, and the university and the Rubber Division set out on a partnership to begin offering the course in the fall of 1962.

Lundberg also thought the perfect textbook for the course was the *Introduction to Rubber Technology* that Morton had edited. The book, originally published in 1959, was a seminal text in the discipline, and the Rubber Division had an important hand in its creation.[85] It emerged due to a growing demand not just from the laboratory scientists and technicians, but also from those involved in rubber sales, service, and purchasing, for a basic primer in rubber technology. As a result, various local Rubber Division groups began sponsoring lectures addressing a wide variety of topics. In 1956, A.E. Juve, then chairman of the Rubber Division, appointed a new publications committee, which concluded that an editor should be appointed to present a survey of the lectures given by these local rubber groups to be used as a textbook. Maurice Morton was offered the position, which he accepted; he published the book in 1959, having also contributed a chapter of his own, entitled an "Introduction to Polymer Chemistry."[86]

With this book as the basic text, Morton and Lundberg agreed that the University of Akron would administer the correspondence course in conjunction with the Rubber Division. All members of the Rubber Division were eligible to sign up for the course, for a $30 fee, and each received a copy of Morton's *Introduction to Rubber Technology*. Participants registered for the course through their local rubber groups. Over a period of fifteen weeks, various topics were covered, and at the end of each week participants were to complete a written assignment. Each assignment was graded and corrected by either Morton or Howard Stephens. Once all the assignments were completed and the course concluded, participants received a certificate.[87]

The response to the first course was overwhelming. Although registration was limited to 300 people, 550 people had signed up. The Rubber Division sent letters to the extra registrants and told them their money could be refunded, or they could be registered for the next course, which would begin in February 1963. In all, only 30 people wanted refunds, and 220 registered for the February class. The educational backgrounds of those who took the first course varied widely. Of the participants, 103 had no college degree, 179 had a bachelor's degree, 12 had a master's degree, and 7 had doctorates. Most worked in rubber laboratories (183), followed by sales (52), plants (34), and offices (32). The registrants came from the following rubber groups: Akron (4), Boston (20), Buffalo (16), Chicago (2), Connecticut (21), Fort Wayne (19), New York (50), Northern California (16), Philadelphia (30), Quebec (4), Rhode Island (15), Southern (57), Washington (8), and nonmembers (39).[88]

Since the Rubber Division's partnership with the University of Akron began more than a half century ago, the scope of its support for the industry has broadened. As a "professional organization dedicated to providing educational programs, technical resources and other vital services for the people associated with rubber and affiliated industries," it achieves its vision through awards, educational courses, library resources, expositions, and publications.[89]

The long history and relationship between the University of Akron and the Rubber Division has benefited both organizations. Today, the Rubber Division's administrative offices and library remain at the university. And while the correspondence course no longer exists in its traditional U.S. Postal Service format, it has evolved into a thriving online educational experience with additional courses added to cover intermediate and advanced rubber technology. Former chairman of the Rubber Division Charles Rader said, "the Division has prospered along with the University of Akron. It has been a long, pleasant and synergistic relationship between the two bodies."[90]

This relationship is now centered on the fourth floor of the Polymer Engineering Academic Center, which opened in 2001. Preparation for its new home in this facility coincided with changes within the Rubber Division, which hired Ed Miller as its permanent executive director in January 2001. An executive director was a new, more professional administrative position and was essential because it was impossible for the chairmen, who also held full-time jobs in industry, to do both jobs, given the growth of the Rubber Division. Under the new administrative structure, the governing body, which drew its members from the Rubber Division, set the goals for the future, and it was the responsibility of the executive director to manage the office and see that these goals were achieved. The structural changes have increased the Rubber Division's efficiency and strengthened its ties to the University of Akron.

The Rubber Division also maintains a strong educational mission. It offers online courses in Basic Rubber Technology, Intermediate Rubber Technology, and Advanced Rubber Technology; it also holds frequent seminars and workshops. The Rubber Division promotes the entire industry and facilitates new relationships through its expositions, technical symposia, business summits, and specialty conferences, that publicize the industry and foster new relationships within it. The three-decade-old Rubber Expo and its associated International Rubber Conference attract over three hundred rubber industry companies, which prepare exhibitions for nearly eight thousand guests. It remains the only large show that focuses exclusively on the rubber industry and rubber-related products.

The library and resource center also remains an important part of the Rubber Division. Librarians like Christopher J. Laursen, manager of the John H. Gifford Memorial Library at the University of Akron, respond to numerous research questions related to the rubber and plastics industry. Laurson also prepares specific literature searches and provides a compilation of Internet resources. The Rubber Division contributes to the library in its own way, sponsoring technical publications in rubber engineering and manufacturing, quality control, extrusion, environmental effects, materials and applications, and general rubber and reference topics. Its *Rubber Chemistry & Technology* remains a leading journal in the field.

Charles Rader continues to foresee a close relationship between the University of Akron and the Rubber Division. Each will encounter problems, which the other will be able to assist; the current guiding philosophy is, "If you work together in a spirit of good will and trust, you can get a whole lot more accomplished."[91]

CONCLUSION—THE END OF AN ERA

Throughout Morton's tenure, there had been many occasions in which the polymer community reflected upon its past, including the initial inauguration of the International Rubber Science Hall of Fame members, the fiftieth anniversary of the first rubber chemistry course, and the twenty-fifth anniversary of the synthetic rubber program. In 1994, the polymer community gathered again to commemorate another important and poignant moment—the passing of the man who orchestrated these events, Maurice Morton.

Between 1952 to 1978, Morton had been at the helm while the university underwent numerous academic changes. These included the establishment of the Ph.D. program in polymer chemistry, the development of the Institute of Rubber Research (and later the Institute of Polymer Science), the expansion and recruitment of eminent faculty members, the evolution of the polymer Ph.D. into an interdisciplinary

Morton is pictured here with his edited book, *Rubber Technology*. It is now in its third edition and is used throughout the world. *Polymer* (Spring 1993).

program, the strengthening of the relationship with the Rubber Division of the American Chemical Society, and the creation of an autonomous Department of Polymer Science. His crowning achievement, however, might be the impact that more than thirteen hundred master's and doctoral students who graduated from the program, since Morton began offering the Ph.D. in polymer chemistry have had on the field. These students have fanned out across the world and are a key factor in maintaining the Akron polymer program's excellent reputation.

Morton could not have been so successful without his partnership with President Norman Auburn. The two worked tirelessly together for many years, until Auburn stepped down as head of the university in 1971. He remained a visible symbol of the past in the Akron area until his death in 2003 at the age of ninety-eight. In 1992, the university honored him as one of the school's "founders," placing him in a company with only three others: John Buchtel, his wife Elizabeth Buchtel, and Park R. Kolbe (school president from 1912 to 1925).[92] Auburn masterminded the school's transition from a municipal to a state school and created several colleges at the university, such as law and nursing.[93] But it is his administrative collaboration with Morton and his name on the Auburn Science and Engineering Center for which he is best remembered by the polymer community.

The administrative achievements of these two men sometimes overshadow Morton's own research and his progress as a polymer scientist. While his success as an administrator is evident, Morton himself had a distinguished career and received many honors from colleagues in his beloved polymer field.[94] By the end of his career, he had authored more than 120 journal articles, written four books, and held four patents. *Introduction to Rubber Technology*, now in its third edition, is still used as a textbook throughout the world. The Plastics and Rubber Institute of London honored him in 1978 with its Colwyn Medal, the British Rubber Industry's most prestigious honor. In 1985, the American Chemical Society awarded him the Rubber Division's Charles Goodyear Medal for his life's work in polymers and synthetic rubber. He won the American Chemical Society's Paul J. Flory Polymer Education Award in 1988. Two years later, he was honored with the Medaille de la Ville de Paris.

The highest recognition bestowed upon Morton by the University of Akron was a renaming of its polymer science program. In 1993, the Institute of Polymer Science changed its name to honor its founder, becoming the Maurice Morton Institute of Polymer Science. Morton himself celebrated his eightieth birthday at the time of the renaming.[95] Nearly two hundred people attended the festivities at the Quaker Square Hilton. An important measure of Morton's success, said Darrell Reneker the director of the Institute of Polymer Science, was a map of the world with markers on it indicating where graduates of the University of Akron's polymer program are working. There were markers on every continent, and Reneker said this demonstrated Morton's "tremendous influence on the polymer field."[96] University of Akron Provost Mark Auburn, son of the president with whom Morton established the Ph.D. program in polymer chemistry, said, "It was a small beginning in 1956, but without it we would not be the comprehensive research institution we are today . . . [It has been] central to the whole life of Akron."[97] A year later, Morton passed away and was elected as the next honoree enshrined in the International Rubber Science Hall of Fame.

The polymer community gathered on April 23, 1994, in the auditorium of the Goodyear Polymer Center to remember a polymer pioneer. His colleagues tried to express what Morton meant to them individually and also to the university. Don McIntyre, head of the Department of Polymer Science, said, "We, the faculty, knew him as a colleague and a leader. In both roles he showed his keen scientific mind, his flair for organizational breakthroughs, and his ever present skill in being straightforward in his teaching, in his research, and in his leadership."[98] R.J. Chang, a principal engineer at Raychem, also expressed his sympathy and respect for Morton, saying, "I shall always remember his tireless spirit and superb vision that laid the foundation of the finest polymer research institution in the world."[99] Frank N. Kelley, his successor, said, "His work is still benefiting the college, its alumni, and the polymer industry. He will be missed."[100] Kelley concluded, "His death marks the end of a very significant era."[101] It would fall upon Kelley's shoulders to shape the next one.

5

THE SUMMIT OF POLYMER VALLEY

1978–1991

IF CHARLES DICKENS COULD HAVE TRAVELED THROUGH time to Akron, Ohio, in the late 1970s, he might have been inspired to write a sequel to his *Tale of Two Cities*. The same language he used to describe Paris and London in the eighteenth century—"the spring of hope . . . the winter of despair"—could have just as easily been applied to the once thriving rubber capital. The tire industry was moving out of town, and it left in its wake mass unemployment, cold smokestacks, and vacant and demolished buildings. When those factories had been active, Akron was a thriving industrial center, the pungent smell of rubber its defining characteristic. Though its absence perhaps meant cleaner air, it also reflected a quiet emptiness that signified the loss of the region's identity. It was a "winter of despair" within the city. The Rubber Capital of the World was no more.

As rubber workers and many of their employers moved out of town, polymers represented a "spring of hope" from the vantage point of the University of Akron. Civic leaders and university officials came to believe that polymers offered a "cure for northeast Ohio's blue-collar blues."[1] Two men, William Muse (the university's president) and Frank Kelley (director of the Institute of Polymer Science) worked together to solidify the program's reputation as "world-class" through several key initiatives that attempted to advance the national standing of the polymer program and redefine the entire region as a prosperous "Polymer Valley." Dale Allen, editor of the *Akron Beacon Journal,* wrote that the polymer program and its facilities at the University of Akron were "symbolic in that to some degree it represents a rising of the polymer phoenix from the ashes of the rubber industry in Akron."[2] By 1991, the university had unveiled the Polymer Science Building (it became the Goodyear Polymer Center in 1998 and will be referred to by that name in this chapter) as the home of its newly-formed College of Polymer Science and Polymer Engineering. Perched overlooking the city of Akron, its mirrored windows radiated its own light from within and at night became a glowing beacon of new life on the skyline.

The city's duality—the remnant of rubber's past and the promise of polymer's future—was apparent even to outsiders. Stephen Franklin wrote in the *Chicago Tribune* about this "confusing vision" of Akron. He described the new "glass-sheathed . . . gleaming" polymer building as representative of Akron's high-tech future. He then contrasted that image with the reality of the city and the skeletal remains of its rubber industry. Walking around Akron, Franklin discovered that a "gloomy hush haunts empty stores in the few downtown streets, and wind whistles by long-shuttered, red-brick plants."[3] What, he wondered, was the proper vision of Akron?

Akron was at a crossroads, and the university was also undergoing a period of transition. In 1978, the polymer program experienced a change in leadership, beginning with Maurice Morton's retirement and Frank Kelley's return to campus. The completion of the Goodyear Polymer Center in 1991 marks an end to that transitional period. The fourteen years in between were a troubled time for the city and a period of great opportunity for the university. Could an academic polymer program descend from its ivory tower to become an economic catalyst for local industry? The sights were set high, but so were the stakes, and the cooperative efforts necessary to establish Polymer Valley, including the formation of the Edison Polymer Innovation Corporation (EPIC), the creation of a new College of Polymer Science and Polymer Engineering, and the erection of the new Goodyear Polymer Center. These efforts and the new glowing building stood in Dickensian contrast to the old world of the rubber workers. Just as rubber pioneers like Seiberling, Goodrich, Firestone, and O'Neil defined Akron one century earlier, perhaps the university's polymer program could forge a new image of the city. In striving to achieve this transformation, the University of Akron became an intellectual summit in the attempt to create a Polymer Valley.

1978—A TIME OF TRANSITION

In her memoir *Gum-Dipped*, Joyce Dyer, the daughter of a Firestone rubber worker, recalled 1978 as a devastating year. The company meant everything to her family and was a constant presence in their lives. Firestone's main plant was so close to her house that its large red neon sign, which radiated the company letters for all of Akron to see, flooded into her bathroom window and onto her pillow when she was a girl. She literally slept under its glow, and she said it "became the nightlight of my youth." This symbol of comfort and safety was demolished with that building and its neon sign in 1978. Her father was out of work, and one of the defining images of her childhood was gone. Dyer wrote, "Over and over we watched the factories turn into empty buildings, the bricks and windows grow darker and more bleak." As for the Building Number 2, the main plant, "A wrecking crew battled its steel and concrete for two years before the building finally fell, a winter sacrifice."[4] The darkness descended for many of Akron's rubber workers. Could the city create a new beacon to glow in the night sky?

The year 1978 marked a critical moment in the history of Akron. In their compelling account of the Akron rubber industry, *Wheels of Fortune*, journalists Steve Love and David Giffels argued that the story of the demise of the Rubber Capital of the World had all the elements of a classic Greek tragedy.[5] "By the end of 1978," they wrote, "the profession that had defined Akron since early in the century—the building of car tires—was virtually gone. . . . Akron lost more in 1978 than perhaps any other year."[6] It was a dire year for rubber companies, both large and small, in Akron. In March, Goodyear's key plant shut down, and twelve hundred workers lost their jobs. In June, Firestone announced the closing of one of its largest tire producing plants, and another twelve hundred workers lost their jobs. In August, Firestone closed yet another plant, and thirteen hundred men and women became unemployed. That same month Mansfield Tire, located about an hour outside Akron, stopped production and displaced seven hundred and fifty workers. In November, Mohawk Rubber announced its impending closure and told over three hundred employees that they needed to start looking for other work. In December, General Tire announced it would have to suspend operations at its tire plant. It asked its workers to take a pay cut in the hope that the company could modernize its plant, but the concessions were for naught, and the plant was never built. Month by month, through 1978, Akron's tire heritage was dismantled.

Downtown Akron was also experiencing problems. The Associated Press reported that in Akron "the pain of transition is apparent," with a city of dilapidated buildings and a large, empty department store.[7] Polsky's Main Street department store had been a centerpiece of downtown Akron for ninety-three years, but it too closed its doors in 1978.[8] With rubber laborers out of work, the once-thriving city was dark and its economy on the edge of collapse. Chrissie Hynde, lead singer for the rock group the Pretenders, wrote a song about her hometown in "My City Was Gone": "There was no downtown . . . My city had been pulled down."

The rubber industry was in a period of transition—out of Akron. By 1978, according to one Goodyear history, "Akron symbolized the past."[9] There were several reasons for the loss. The first was due to technological obsolescence: the Akron plants still produced the older bias-ply tires, despite the superiority of new radial tires. Second, the Akron rubber industry had old, outdated factories with a multistory design in an age where the most efficient plants were single story. Finally, labor-management disagreements weakened the companies both groups cared so much about. The rubber workers themselves, in attempting to fight for their jobs and secure better working conditions through strikes and labor negotiations, unknowingly played a key role in driving work out of Akron. The management that had so long enjoyed prosperity became complacent, failed to win government incentives, and let its factories become obsolete. The major companies either moved their headquarters out of Akron, split up into smaller operating units, or were purchased by foreign firms.

In the 1980s, the forces of globalization, perhaps more than anything else, continued to transform the face of Akron's industry. The development of radial tires was in part responsible for this shift. These were first developed by the French company Michelin in 1946 and offered numerous benefits over the bias tire both in overall life and increased fuel economy. Four years after their introduction, they became standard on all new French cars, and the rest of Europe soon followed. The United States was much slower to adopt the new technology. By 1975, nearly 60 percent of all United States tires were still bias, but the American radial revolution was about to occur. According to one account, "The impact was immediate and profound. When the shakeout sparked by radials ended, the world's tire-making borders had been completely redrawn."[10] In 1987, Germany's Continental AG purchased General Tire. In 1988, Japan's Bridgestone merged with Firestone, with Harvey Firestone's name now being listed second. That same year, the Italian company Pirelli acquired Armstrong Tire. In 1989, Japan's Yokohama purchased Mohawk, and the French company Michelin bought Uniroyal Goodrich. The industrial face of Akron was changed forever.

While the rubber industry experienced despair in 1978, that year was a time of quiet optimism at the University of Akron. It was not without change—Maurice Morton stepped down as head of the polymer program, and Frank Kelley returned to advance a polymer program with a strong international reputation. Kelley was well aware of the problems facing the rubber companies in Akron. Born in Akron's North Hill neighborhood on January 19, 1935, he became a third-generation Goodyear employee after graduating high school. He started in 1952 as a filing clerk in the record storage department, where he spent his days throwing out old records and putting new ones into the vaults. He soon became disenchanted with the job. His father, John Kelley, a Goodyear accountant, made a suggestion. He told his son he should study rubber chemistry at the University of Akron. Kelley took his father's advice, pooled his limited savings with some scholarship money, and enrolled in college—the first member of his family to do so. But the excitement over beginning this new phase of his life was tempered with tragedy when he learned that his father was dying from cancer; he passed away during Kelley's junior year.[11]

Kelley combined his studies with factory work at Goodyear. As a member of its "Flying Squadron," he was assigned to a variety of different jobs in the plant.[12] Kelley learned to appreciate all aspects of the tire-making process. He soon found he preferred the intellectual challenges of rubber science to those of the manual labor of building tires, and he pursued a degree in chemistry. He also began taking graduate-level polymer courses while still working on his bachelor's degree. These formative years made a lasting impression on Kelley. As a student, he was fortunate to have worked with many of the people who shaped the polymer discipline: he served as Ralph Milkovich's assistant; he observed the old master, G.S. Whitby, in the laboratory; and he interacted closely with a vibrant and youthful Maurice Morton. James Berry was Kelley's advisor during his master's work, supervising his research on ways to make model networks from polymer molecules. The work paved the way to a summer job at Goodyear research, which led to a Goodyear Fellowship and the beginning of Kelley's doctoral work.

Fred Bueche served as Kelley's Ph.D. advisor and offered him what he considered a "risky" and "dramatic" choice of a dissertation relating to the adhesion between cured and uncured polymers. As a side project, Kelley also studied the effect of glass transition temperature (the temperature below which molecules have little mobility) on molecular weight. Kelley determined that he could calculate the depression of the glass transition temperature of a polymer with plasticizers or liquid. This came to be known as the Kelley-Bueche Equation. As a graduate student in 1960, Kelley and some of his colleagues were profiled in the *Akron Beacon Journal* as the next generation of those "Probing Rubber's Secrets."[13] Included in this group was Irja Piirma, who later became a full professor of polymer science at the university and eventually head of the Department of Polymer Science. She researched the relationship between the pull strength of polymers and how their molecular chains were assembled. Another student profiled was Nissim Calderon, later vice president of research at Goodyear. He worked to understand how the formation of long-chain molecules occurred.

As an advisor and mentor, Bueche became a role model for Kelley's own approach to graduate advising later in his career. Kelley was always impressed with Bueche's simple and elegant approaches to experimentation. His observations often required nothing more than a stop watch, a ring stand, and a microscope. Yet despite the very basic set-up, he made important contributions to polymer physics. Bueche left Akron in 1961 and was replaced by Alan Gent, just as Kelley was finishing his Ph.D.—a member of the third Ph.D. class to emerge from Akron. When Kelley received his diploma, President Norman Auburn expressed a desire that he might one day return. Auburn would get his wish but had to wait seventeen years.

Even if he had wanted to stay in Akron, Kelley had other obligations to fulfill. As an undergraduate, he was a member of the ROTC and was, therefore, required to commit to three years of service in the air force after graduation. After a short three-month stint at Union Carbide, Kelley was assigned to the Rocket Propulsion Laboratory at Edwards Air Force Base in California. In late 1961, he and his wife drove across country to the dry lake beds in the desert, where the new rocket laboratory was located. This was just four years after the Sputnik launch, and the United States was allocating tremendous resources to "catching up" with the Soviets. Edwards Air Force Base had, therefore, changed its primary mission from rocket engine testing and evaluation to rocket research and development. Although Kelley enjoyed the research and managerial aspects of his new position, the sand fleas that infested his first housing unit were something he and his wife never got used to.

Kelley was one of the few Ph.D.s on the base. His job allowed him the opportunity to keep his own laboratory, which he used to research the mechanical properties of solid-propellants. He completed his military service in 1964 as a lieutenant but remained as a civilian, becoming the branch chief of propellant development in 1966. In 1971, he was promoted to chief scientist, the highest ranking civilian position in the laboratory. Although the research was fulfilling and he steadily advanced up the chain of command at the base, he often thought about moving into an academic position in a university. Ultimately, he had one career goal. He said he wanted to "reestablish my credibility as an academic, and maybe I could get a job in Akron one day."[14] For the next few years, however, he remained with the military and in 1973 moved back to Ohio, where he became the chief scientist of the Air Force Materials Laboratory at Wright-Patterson Air Force Base. Here he had a jet at his disposal, which he used to attend his frequent Pentagon briefings and American Chemical Society meetings. In 1977, he was promoted to director of the laboratory. One year later, at the 1978 American Chemical Society meeting in New Orleans, Maurice Morton and several members of the search committee replacing Morton approached Kelley and asked if he might be interested in becoming director of the Institute of Polymer Science at the University of Akron. Kelley eagerly applied, and the search committee concluded he was the best candidate for the job. In September 1978, he returned home to Akron to take over the helm of the Institute of Polymer Science.[15]

Kelley would face a difficult period of transition. Several questions loomed as he moved into Morton's office in the Auburn Science and Engineering Center. Funding remained essential to supporting the faculty research programs, and the yearly budget was just over a million dollars. This included $280,209 from general funding, $739,346 from grants, $100,055 from fellowships, and $43,343 from applied research services. Could Kelley develop new funding for the polymer program? He was taking over from the dynamic, charismatic, and forceful leader who developed the Institute of Polymer Science into a position of international prominence. Could Kelley maintain Morton's momentum given his vastly different personal and managerial style? Would he be able to find new directions in which to grow the institute? There were other questions as well. The year he returned, 1978, had been devastating for the rubber industry. There had always been a recipro-

Twenty-fifth anniversary cartoon of the Ph.D. program in polymer science. Kelley was depicted in the background, with an incorrect spelling of his name (1981). *Akron Beacon Journal.*

cal relationship between industry and the Akron campus. Could the polymer program at the university serve as a lightning rod to stimulate economic growth and revitalize the region?

Kelley's transition turned out to be more difficult than he had imagined. One important administrative change occurred within the polymer program when Morton stepped down. Morton had held the position of institute director, as well as that of head of the Department of Polymer Science. After he retired, the two positions were split, with faculty members rotating through the leadership positions in the department, while the institute director remained a permanent post. The institute's mission was research, while the department's was instructional. The responsibilities of the Institute of Polymer Science and its director were to oversee and manage the research activities, the support staff, and the program facilities. The responsibilities of the Department of Polymer Science and its chair were to assure financial support, recruit and assign faculty members, recruit and advise students, schedule classes, update the curriculum, and assess academic performance. While Morton unified these important roles under his own control, splitting them brought some initial difficulties.

Frank Kelley in his laboratory in 1985. He was an active researcher and teacher during his entire tenure at the University of Akron, and had ten graduate and two postdoctoral students in his group. *The University of Akron.*

Kelley later reflected, "It was a really trying, difficult thing for me. I thought it was going to be very easy. I had all of this management training. I had all of this strategic planning activity behind me. I had interacted with the highest levels of government, as well as with academics and Nobel laureates."[16] But Kelley's plans to make changes to the institute met with "tremendous resistance." Morton had built a program based upon his own power and strength. Kelley described it as "seeded with arrogance and brutality in some ways." Everyone admired Morton's abilities and achievements, but his heavy-handed leadership style often strained interpersonal relationships between the director and his faculty. Alan Gent recalled that the university was in many ways a totalitarian régime during the Norman Auburn years. Morton himself, though a dynamic leader, "couldn't resist a fight," and he commonly responded to issues "through aggression."[17] When Morton stepped down, he left behind faculty who resented the administrators. The Morton era was coming to a close, and Kelley was ready for the challenge of his life.

A PLAN FOR THE '80S

While decades are, in reality, arbitrary moments in time, they often serve as benchmarks to plan for the future. A new decade offers the opportunity for new resolutions for change and advancement. The president of the University of Akron felt that such a moment was at hand in 1980. Dominic J. Guzzetta enjoyed a long career at the university. He had come to the school in 1954 and held a variety of academic posts, including provost, before becoming Auburn's successor as president in 1971. He viewed 1980 as a "threshold between one exciting decade and another." The past ten years had seen new buildings shape the campus, among them the E.J. Thomas Performing Arts Hall. With its dramatic slanted architectural lines, it became an aesthetic symbol and a living monument to progressive ideals within the city. Guzzetta wrote, "what has been built in the past . . . can be built upon further."[18]

Before embarking on major plans for the future, the university had another important milestone to celebrate—the twenty-fifth anniversary of the Institute of Polymer Science and the establishment of its

Ph.D. program.[19] With festivities held over a two-day period in September 1981, four distinguished guests came to discuss the state of polymer science. Herman F. Mark, founder of the Polymer Research Institute at Brooklyn Polytechnic, spoke on "Polymer Science: Aging Roots—Sprouting Branches." Maurice Morton, former director of the Institute of Polymer Science, gave his presentation on "Controlling the Macromolecule." Carl S. Marvel, from the University of Arizona, discussed "The Organic Chemist's Contribution to Polymer Chemistry." Leonard Mullins, director of the Malaysian Rubber Producers' Research Association, spoke on "Understanding Rubberlike Materials."[20] The event focused the world's polymer attention on Akron and for a moment at least overshadowed the troubling news emerging daily from its tire industry.

The institute had much to celebrate. The polymer program had been responsible for graduating 193 Ph.D. students since 1959—a number the university considered a "world record."[21] Thirteen faculty members on staff were making important contributions to the field. A snapshot of this group in 1981 at the twenty-fifth anniversary included some distinguished individuals. Lewis J. Fetters was the first recipient of the Creative Polymer Chemistry Award from the American Chemical Society, and a participant in the first China-U.S. Bilateral Symposium on Polymer Chemistry and Physics in 1979. John E. Frederick had been the chairman of a key symposium on the use of lasers for research in polymer science for the American Chemical Society. Alan N. Gent had won numerous awards in the field, and was also serving as the dean of graduate studies for the university. Gary R. Hamed was the author of eleven scientific papers (one of which won the 1980 Best Paper Award at the American Chemical Society) and held two patents. James Harwood was considered the University of Akron's "Outstanding Researcher" in 1980. He had also been one of nine American delegates at the United States and China seminar on copolymers. Frank Kelley was serving as the director of the Institute of Polymer Science. Joseph P. Kennedy had been chosen as an honorary fellow of the Society of Polymer Science of India in 1980. In 1965, he was Japan's first industrial guest professor of polymer chemistry; he had published more than 140 scientific papers and held over twenty-four patents. Donald McIntyre was the author of more than fifty publications and had been a visiting scientist at the Centre de Recherche sur les Macromolecules in France. Eberhard A. Meinecke specialized in viscoelasticity, optical properties, and processing and had over fifty publications in the polymer field. Howard L. Stephens headed the Department of Polymer Science at the time (he served from 1978 to 1982), published widely, and gave presentations throughout government and industry. Irja Piirma had been the first woman to earn a Ph.D. at the University of Akron in 1960. She remained on the faculty as an active researcher and publisher and also served as department head from 1982 to 1985. Charles W. Wilson III specialized in polymer research with nuclear magnetic resonance, and he had more than twenty publications at that point in his career. Maurice Morton remained as an emeritus faculty member and continued to do research and serve on the editorial board of the *Journal of Polymer Science* and the *Journal of Applied Polymer Science*.

The twenty-fifth anniversary celebrations attracted the wider polymer community to Akron for the occasion. After it was over, Herman Mark called it a "truly memorable meeting."[22] Although Paul Flory was unable to attend, he wrote a letter congratulating the entire university for its "pioneering step in the establishment . . . of a major center of polymer research and education."[23] The celebration also reinforced the ties between the university and local rubber industry leaders. Frank Kelley ensured that vice presidents and directors of research from the four major rubber companies were honorary members of the planning committee. This included Carl Snyder from Goodyear, S.L. Aggarwal from General Tire, A.G. Altenau from Firestone, and Robert Fawcett from B.F. Goodrich. Kelley kept in contact with all of them, writing personal letters expressing his desire to strengthen the academic/industrial connections. He wrote, "I hope also that this present purpose will serve as a good beginning for possible continued associations."[24]

With the celebrations complete, the university could refocus on its plans for the future. In an effort to take the institute to the next level, Kelley and his staff developed a "Plan for the '80s," which projected the organizational and programmatic needs for the next decade.[25] Kelley wanted to modernize the program, update the curriculum, and move it beyond rubber itself. Underlying this plan was the assumption that polymer science would continue to be of importance both to the country and the university. Another assumption was that the climate of research would be such that federal funding would become tighter, while the importance of academic-industrial partnerships would be elevated. Within this academic climate, Kelley believed his institute would have to make numerous choices to ensure its place at the leading edge of polymer research. In laying out a series of goals, Kelley preferred changes that were the "outcome of proactive rather than reactive strategies." Dominic Guzzetta retired from the university in 1983, and it fell to a partnership between Frank Kelley and new president William Muse to determine the future of the polymer program at the university and ensure that its new goals would be achieved.

A close relationship between university president and head of the polymer program had been of vital importance in its past. The bond between Maurice Morton and Norman Auburn had been absolutely essential in establishing the prominence of the Institute of Polymer Science (and the Institute of Rubber Research before it) and the Ph.D. program in polymer chemistry. For the polymer program to reach the next level, a new partnership between leaders at the university was necessary. When Muse arrived in 1984, those involved in the program wondered, could he appreciate the importance of polymers? The question was soon answered as the new president demonstrated that he shared Kelley's vision for a bold new future supported by the strength of the polymer program.

Although this was his first time serving as a university president, Muse had a long track record in academia. He was born in 1939 in Mississippi and graduated from Northwestern State University, with a degree in accounting in 1960. He then attended the University of Arkansas, earning his M.B.A. in 1961 and a Ph.D. in business administration in 1966. He remained in academia, becoming a business school professor first at Georgia Tech and then at Ohio University. In 1970, at the age of thirty-one, he became one of the youngest deans of a business college in the United States. He held this position at three successive universities: Appalachian State University (1970–73), the University of Nebraska (1973–79), and Texas A&M University (1979–83). From 1983 to 1984, he served as vice chancellor for academic programs at Texas

A&M, and at that point he began considering a position as a university president.

Although he had spent five years in Athens, Ohio, Muse had never visited Akron. During his first trip, while he was being considered as Guzzetta's replacement, Muse observed the extent of Akron's difficulties firsthand. The loss of tire manufacturing had an "economic as well as a psychological impact on the city." He found that many of those he met had a "defeatist attitude," but he believed new mayor Tom Sawyer was a "positive force" who could take the city in a new direction. Upon returning to Texas, Muse concluded that his experience as a business school dean and the potential partnership with the city through its mayor represented a tremendous opportunity. With a strong partnership, he thought that the university could "have a significant impact on the city and the region."[26] Muse became the University of Akron's twelfth president in 1984. At his investiture ceremony, he clearly stated his plan for more closely aligning the university with the city, proclaiming, "We can be a significant catalyst in the revitalization of northeastern Ohio, economically and culturally."[27]

Unfortunately, the historic divide between the city and the university had grown acute during the previous decade. The city of Akron and the university often had a troubled relationship since the school ceased being a municipal institution in 1967. But Muse and Mayor Tom Sawyer wanted to improve their association significantly. They strove to establish a "symbiotic relationship," believing the "two dissimilar organisms" might live more closely for mutual benefit.[28] Muse thought the university should and could become one of the city's greatest assets. One of the partnership's central goals (which extended back to Guzzetta's presidency) was the "Span-the-Tracks" program. Several rows of railroad tracks formed a literal boundary between the university and the city. Civic and academic leaders believed that if this mental and physical boundary could be bridged, then the two could work more closely together. This was, in part, achieved by keeping the lines of communication constantly open. Muse, along with Donald L. Bowles, vice president for administrative services, and Roger Ryan, the university's architect, met with the mayor and his staff on a regular basis. One key strategy in establishing this relationship with the city was through the polymer science program.

When he first came to town, Muse knew very little about polymer science. But he said that after sitting down with Frank Kelley he was immediately "impressed with the strength of the program." Furthermore, he said he was "impressed with its relevance to the city and the region and felt that it was a program that should be emphasized."[29] Muse believed in a strategy of matching the university's mission to its market. The university should cultivate its strengths in areas most useful to the region it served. Muse soon realized that "There was no doubt that for the Akron market and region, the most relevant program in terms of future growth and the ability to attract industry was the area of polymers."[30]

One other reason Muse believed the polymer science program should have more resources committed to it was something he called "Peaks of Excellence." He said that if a university wanted to improve its overall status and become more highly regarded in the business and academic community, it needed to have some programs that were nationally ranked at the top of their fields. When outsiders looked at a university as a whole, they would first see these academic "peaks" and would be more likely to regard the entire school more favorably. When Muse first came to Akron, there were few programs capable of achieving national rankings, but polymer science already had a strong reputation on which to build. What he thought was strange was that the university had not committed more funds to helping the program achieve wider notoriety. Muse said, "It was a little mysterious to me that it was kind of buried within the University." Indeed, polymer science was just another department in the Auburn Science and Engineering Center.

Muse and Kelley knew they needed to work closely together to achieve their ambitious goals for the university and the region. They created a roadmap, beginning with the goals Kelley already had in place, and they devised a true master plan to guide them over the coming years. The master plan included developing Ohio Eminent Scholars, establishing privately endowed chairs, building a relationship with Case Western Reserve University's polymer program, providing more resources to polymer engineering, consolidating all polymer departments and programs into one new college, and constructing a polymer science building.

The plan for the eighties was in place. Muse began promoting polymer science from the outside as a "peak of excellence" and reinforcing the discipline's link to the local economy. Likewise, Kelley worked to gain more visibility and resources for polymer science and establish a cooperative venture between industry and academia to create a "Polymer Valley." Their efforts would culminate in a dramatic new building that became a literal "peak of excellence" on the Akron skyline.

AN EPIC PLAN

In 1971, Don Hoeffler published a series of influential articles describing an increasing concentration of computer and semiconductor companies located in the San Francisco Bay Area. He proposed the name "Silicon Valley, U.S.A." to describe an economic region consisting of start-up companies and their relationship to Stanford University. California's Silicon Valley, along with the Research Triangle in North Carolina and Route 128 in Boston (both of which emerged in the 1950s), exemplified a new research model.[31] The model called for establishing relationships between universities and industries, linking them to training facilities, and providing sources of funding. In northeast Ohio, politicians, academics, and industry leaders all agreed they had the basic ingredients at their disposal to create a similar research area that revolved around polymers. The northeast Ohio "polymer prophecy" first appeared in 1979, with a study funded by the Economic Development Administration that explored ways in which Akron could diversify its economy. The study was never officially released to the public and, after meetings with the "Big Four" rubber companies, a comprehensive plan on how to capitalize on these suggestions fell apart.[32] It appeared that Akron's polymer community would be unable to follow the silicon footsteps in California.

Two years later, in 1981, these polymer dreams resurfaced in a plan called Summit 21. Its goal was to examine ways in which the region could reinvent itself after the downturn in the automobile industry. It concluded that the polymer research taking place in Akron was a "dormant resource" and with the right cooperation and funding, Akron could be cultivated into a new Silicon Valley.[33] Journalists immediately latched onto the idea, telling readers to "Think of the Silicon Valley in California. Then Substitute the word 'polymer' for silicon." One reporter said that this had the "makings of an economic boom."[34]

At the same time the Summit 21 report was published, a similar independent study in Cleveland came to an identical conclusion. A survey of seventy-five top executives in the Cleveland area assessed support for an automotive technology center.[35] The surprising result was that few of them were interested in cars. Instead, nearly all of them thought polymers should be the regional economic focus for the future. Because of the strength of the polymer program at Case Western Reserve University, polymers presented Cleveland with its best chance for economic revival as well. Like Akron, Cleveland, too, had fallen on hard economic times, losing more than ninety thousand manufacturing jobs since the late 1970s. Donald Iannone, the director of business development for the Greater Cleveland Growth Association, said the city was getting serious about polymers. "It's one of the best potential development targets we have," he said. "It's a great opportunity for regional cooperation."[36]

Leaders at the University of Akron and Case took these reports seriously and began their own discussions about the polymer future of Ohio. In 1982, Frank Kelley, James White (the director of the University of Akron's polymer engineering program), and other civic leaders began meeting at Akron Mayor John Ballard's house in Merriman Valley. The meetings had what Kelley recalled as a "simple agenda concerning what might be done to revitalize the economy."[37] They started working with Governor Richard Celeste and his administration to try to attract state funding. They advanced the idea that, with the combined academic forces of the University of Akron and Case, along with proponents from the rubber industry, the entire economy could be revitalized with a polymer focus.

There was a great deal of support for this plan from the local rubber industry. At an Akron Roundtable meeting in 1983, Robert E. Mercer, chairman of the board at Goodyear, thought that a partnership between these two universities would help improve the struggling economy. He said that the University of Akron in particular was a "real asset, not only for the community's progress in education and culture but to its economic progress as well. It can be a strong magnet in attracting technology and service firms and very definitely should have a prominent place in our development and promotion packages."[38]

Further defining the ways in which the polymer program at the University of Akron could positively affect the city was an Akron Regional Development Board's Polymer Task Force report in February 1984. It suggested that cooperation between academia, government, and the private sector could strengthen research and development at the university and in industry, improve the process of technology transfer to small and midsized companies, facilitate polymer research into commercial products, and establish entrepreneurial advice for polymer start-up companies. Members of the group included Kelley and Mayor Ballard, as well as high-ranking officials at Akron rubber companies, local banks, and university professors.[39]

The group needed a name for its new initiative. While some began calling the initiative Polymer Triangle, Polymer Diamond, and Polymer Path, the Akron Regional Development Board selected "Polymer Heights." Ironically, the opposite name caught on with the press—"Polymer Valley." No one is sure where this term originated. Some say that Richard Knight, a European economist who was a guest lecturer at the University of Akron, coined the term, while others credit former university president Dominic Guzzetta. Nevertheless, "Polymer Valley" stuck, even though it was located predominately in Summit County, the highest point in the state.[40] Nobel Prize-winner Paul Flory opposed using the name because he thought it sounded like a simple marketing scheme copying Silicon Valley. Akron had tremendous, original scientific potential in its own right, and in Flory's mind the city did not need to replicate another region's successful research formula.[41] Although Frank Kelley was never a proponent of the name, either, he could not resist the groundswell of enthusiasm from those who used it at every opportunity. President William Muse said he never gave a speech without invoking it at least once. Everyone who began talking about Polymer Valley was convinced that it would transform the region as long as the universities cooperated, industry was supportive, and the government provided funding. Eric Baer (director of the Department of Macromolecular Science at Case Western Reserve University) said, "If we can create that climate, then Polymer Valley will happen."[42]

But others were not as certain. One of the most vocal skeptics was Joe Coates, a "futurist" consultant who performed a study on the Akron polymer industry for Goodyear.[43] "Akron U. is a sound base," he said. "It's a good, classy, sound operation. But it's a far cry from what you'd want to create a new industry."[44] Polymer Valley was a pipe dream and a long shot. But he qualified this by saying that so, too, were the Brooklyn Bridge, television, and penicillin. Skeptics suggested such a challenging undertaking would take time, patience, and hard work in order to achieve its goals, and some questioned whether Akron could develop and support this kind of climate.

The findings of another major polymer study predicted that Coates would be wrong. In early 1985, the city of Akron selected the Center for Urban Studies at the University of Akron to examine the local polymer industry and analyze its polymer potential. The report concluded by saying that Akron had no choice but to look for a new future. The city could not only "take charge of its destiny," it could also choose its own future. In part, the authors believed this was possible because the region still retained private businesses and world-class public institutions (namely the university polymer program) that continued to push the boundaries of polymer research. Success, however, was not certain. The "forging of this chain" required the active participation of all sectors. Furthermore, the authors warned, the "path to polymer prosperity is neither preordained nor straight and there certainly is no returning to the Golden Age."[45] Some criticized the report for not offering more specific recommendations, but it was considered the "most comprehensive study ever conducted on polymers." Goodyear chairman Robert Mercer said he was "very much impressed," while President Muse said the study had "confirmed our feeling that . . . polymers [are] a major economic opportunity.[46]

To announce the report's findings, Akron Mayor Tom Sawyer addressed a group of over two hundred local business leaders and city officials in December 1985 at the Quaker Square Hilton. Displayed behind the podium was a vast array of polymer products, the highlight a bright red Pontiac Fiero. Sawyer used the car, with its polymer body and numerous polymer interior components, to symbolize the polymer future. Sawyer's presentation emphasized the Center for Urban Studies Report's findings that the city could make a successful transition from an economy dependent upon traditional manufacturing to one based upon the "innovative applications of new polymer technology."[47] Akron, according to Sawyer, was "poised at the base of a mountain that could reach heights beyond any reached before."[48]

By 1985, the realization that something special was happening in Akron began to spread through the academic polymer community. William McKnight, chair of the Department of Polymer Science and Engineering at the University of Massachusetts said, "there is a good chance that Akron can pull it off and be the major development center."[49] McKnight thought the Akron area was similar to Boston in that Boston's textile-based industry had left town, like Akron's fleeing rubber industry. In Boston's case, the resurgence was fueled by cooperative research efforts between industry and Harvard, MIT, and the University of Massachusetts. McKnight also thought the Silicon Valley comparisons were inappropriate. Silicon Valley was a very different scenario: it was much easier to build computers in a garage than to create polymers. A computer could be put together relatively quickly with fairly inexpensive components. A polymer laboratory required millions of dollars to establish, and years of work by a committed, doctoral-level research team to produce a commercially viable product. Although it might not become a true polymer replication of Silicon Valley, McKnight thought Akron could achieve many of its ambitious goals. The best opportunity it had for success came from a potential partnership with another leading polymer academic powerhouse—Case Western Reserve University. This partnership took shape through an organization established by the Thomas Edison Program, appropriately named EPIC.

In 1983, the State of Ohio established its Thomas Edison Program as a way to partner academic, industry, and government researchers. Organizations wanting to participate wrote proposals outlining rationales for using state funds to promote academic research and technology transfers to local industry. The University of Akron and Case had first applied to become an Advanced Technology Application Center in Polymers (later the name changed to Edison Technology Center) in April 1984. The proposal's authors stated that the program would "catalyze the creation of the most significant polymer educational, research and development complex in our nation and the world."[50] The team writing the proposal included Frank Kelley (director of the Institute of Polymer Science), James L. White (director of the Polymer Engineering Center), and Dominic Guzzetta (university president), and from Case Western Reserve University, Anne Hiltner (director of the Center for Applied Polymer Research), Jerome B. Lando (chairman of the Department of Macromolecular Science), and David Ragone (university president).

The partnership between Cleveland and Akron was a natural one, even though the two schools competed for students, faculty, and grants. The Case faculty had been teaching polymer science and engineering since the 1930s, and the university had officially established its own Department of Macromolecular Science and Engineering in 1962. In the mid-1980s, it had fifteen full-time faculty on staff, as well as additional associated faculty from other departments. Case offered bachelor's and master's degrees, as well as the Ph.D. in polymer science and polymer engineering. In an effort to improve and enhance relationships with industry, Case formed the Center for Applied Polymer Research (CAPRI) in 1981.[51] Ragone, Case's president, was enthusiastic about the collaboration with the University of Akron. He thought that, with patience and careful planning, "sustained and real growth" was possible—though he warned that Silicon Valley did not emerge overnight.[52]

The Akron/Case proposal was accepted, and in July 1984 Ohio Governor Richard F. Celeste officially announced the newest Edison Technology Center. One month later, the Ohio Controlling Board approved a $4.1 million grant to be used by the University of Akron and Case Western Reserve University.[53] Soon thereafter, Kelley and Baer decided it was time to name their program, and they started brainstorming ideas and acronyms. They knew the name had to include the words Edison, polymer, and corporation. While they talked on the phone, Baer's colleague Anne Hiltner wrote the initial letters of these words on a blackboard. They needed a vowel, so Hiltner suggested the word "innovation" and put an "I" between the "P" and the "C." Baer said the word "EPIC," and Kelley knew that they had found the perfect name—The Edison Polymer Innovation Corporation, or EPIC. They were ready to begin turning Polymer Valley into the polymer "EPICenter" of the world. Governor Celeste called the program a "historic first step" that would greatly improve the state's entire economic future.[54]

Many believed that, through cooperation, funding, dedication, and hard work, EPIC would "provide the magic needed for [an] economic transformation."[55] The initial group of twelve companies grew to eighty by 1988. Member companies paid corporate dues, determined by a sliding scale based upon gross sales. This enabled both small and large organizations to become members. EPIC's administrators used the funds to cover operating costs and to sponsor agreed-upon research.[56] Member organizations then shared in the research's results. EPIC's primary goal was to "develop new materials and research technology that shows promise in rejuvenating the state's economy."[57] George H. Kaull became chairman of EPIC, and Robert J. Fawcett (Vice President of Research and Development at B.F. Goodrich) became president.

The two universities, separated by only an hour's drive, each had their own areas of basic research within EPIC. The University of Akron focused on ways to improve the performance of fabricated parts, thermoplastic processing technology, and advanced measurement methods. Case coordinated its work through its Department of Macromolecular Science and CAPRI. It directed research on polymer blends, polymer alloys, membranes, coatings, and lightweight composites. Anne Hiltner, director of the CAPRI program, said, "the new uses of plastics are in very high-performance or specialty areas. That's where the challenges are. The materials have to be designed for the use one has in mind."[58] *U.S. News and World Report* ranked both polymer programs in the top three in the United States. The coordinated efforts of EPIC became a vital partnership for the region. Frank Kelley said in 1985, "The word is out that a connection in the Akron-Cleveland axis is essential to any polymer-related undertaking."[59]

During the first several years, EPIC had some important successes. Major corporations such as B.F. Goodrich, Firestone, and Goodyear were contributing, along with a total of four hundred researchers from eighty smaller companies.[60] Their work led to some specific early products. EPIC patented a new bandage that promoted faster healing, and B.F. Goodrich purchased a licensing agreement to develop it. EPIC also played a key role in helping to form two start-up companies—Polymer Container Corporation in Akron and NanoFilm in Valley View. Polymer Container Corporation was established in 1986 with the idea to develop a better aerosol can, using a balloonlike rubber for self-pressurizing.[61] NanoFilm, established in 1985, developed thin coating to protect optical surfaces from dirt, dust, and oils from skin. By 1989, the *New York Times* reported on the EPIC successes with a cartoon that showed a book entitled *Treatise on Polymers* wearing a graduation cap shaking hands with a bottle dressed in a suit coat and holding a briefcase.[62] This unprecedented cooperative shared research was seen as EPIC's key contribution and as a lesson in how to compete with foreign markets.

Changes were in store for EPIC at the end of the eighties. In 1989, E. Chuck Galloway replaced Robert J. Fawcett as EPIC's president at its Broadview Heights headquarters. This location was aptly placed almost exactly equidistant between the two universities. At that time, Galloway reflected upon EPIC's brief history and pointed to its effort in the creation or retention of 520 jobs as evidence that the program was working. Although this was a relatively small number, Galloway hoped to see increases in the future. Goals for the coming decade included more patents and licensing agreements and more corporate members, up from its level of eighty. A stronger marketing program and offers of support to small businesses could accomplish this. Galloway's final goal was to do more with less state funding. In 1991 and 1992, the state planned to grant EPIC $3.7 million for long-term economic development. EPIC was at an important adolescent crossroads. Galloway predicted, "The next three to five years are going to be the time to make this pay off."[63] Would the EPIC dreams and Galloway's goals realized or would the enterprise fail, as some suggested in 1985, due to "changing political winds?"[64] The answer to this question would come in the 1990s.

CONSTRUCTING A COLLEGE

Former University of Akron President Norman P. Auburn once had a saying, "Nothing is more difficult to move than a cemetery—except a University faculty."[65] Despite the challenges, Frank Kelley and William Muse decided to do just that and move not one, but two sets of faculty at the same time into a newly formed college. The move was necessitated by the emergence of polymer engineering as its own discipline, separate from polymer science. Throughout the 1980s, the emphasis on polymer engineering increased at the University of Akron. In 1983, polymer engineering became its own department under the leadership of James White. For its first few years, the Department of Polymer Engineering existed under the control of the College of Engineering. While this seemed like an obvious choice, it was complicated by the fact that the Department of Polymer Science was part of the College of Arts and Sciences. President Muse thought polymer science and polymer engineering should be brought together and given "a much more prominent position within the academic programs of the University."[66] The way to do this was to actually create a new college, and Muse proposed this plan for the consideration of the Board of Trustees.

Although the board was receptive, existing university colleges often resent being asked to relinquish control of their programs and resources. The question was how to reorganize the university so the Department of Polymer Science could leave the College of Arts and Sciences and the Department of Polymer Engineering could leave the College of Engineering. Claibourne E. Griffin, the Arts and Sciences dean, and Louis A. Hill, Jr., the College of Engineering dean, were for the most part agreeable to the idea. H. James Harwood, head of the Department of Polymer Science during the period when the new college was established, said this was not an easy task because polymer science was much larger than polymer engineering at that time. As a result, he said, the "people in polymer engineering were suspicious that we would overrun them."[67]

In early December 1987, the University of Akron officially asked its Board of Trustees to approve the plan for the new college. This new administrative structure was important, according to Kelley, because

"Our polymer programs are growing rapidly and they needed—for the sake of efficiency—a clearly defined structure that was less confusing."[68] One issue to be resolved was the name. While some wanted simply "College of Polymer Science and Engineering," Dean Hill believed the word "polymer" had to appear paired with both science and engineering. Otherwise, some might conclude that the college had subsumed the entire College of Engineering. Therefore, it became the College of Polymer Science and Polymer Engineering.

Ben Ammons, the chairman of the Board of Trustees, had no objection to the plan and believed the consolidation would be an important, positive step for the university. On July 1, 1988, the University of Akron established its new college and began looking for a new dean to head it. After a national search, the committee offered the position to Frank Kelley. It became the first college in the United States with the exclusive focus on graduate training (master's and Ph.D. students) for polymer scientists and polymer engineers. The college combined the resources and controlled the administration of the Institute of Polymer Science (Darrell H. Reneker as director), the Department of Polymer Science (Donald McIntyre as head), and the Department and Institute of Polymer Engineering (James L. White as both head and director). The faculty of the new college welcomed the change. Stephen Z.D. Cheng, who had joined the Department of Polymer Science a year before the college was established, called its formation a "critical milestone" for the university.[69] Joseph P. Kennedy, who had been a member of the faculty since 1970, called the new college "unprecedented in the United States and perhaps in the whole world."[70]

One of Kelley's first tasks was to fill the position of associate dean. In the midst of the search, he attended a University of Akron 1958 alumni dinner and sat next to Donald Bowles and his wife Joan. Bowles had been working at the university for nearly thirty years, holding a variety of positions, including vice president for administrative services. At the time, he was President Muse's executive assistant and—equally important, he might have said—the shortstop on Muse's Marauders softball team. Kelley mentioned that he was looking to fill this position, and Bowles told him it would be his dream job. Kelley said right then that, if Bowles wanted it, the job was his. Later that night, after talking with

In 1990, Dr. Darrell H. Reneker was the director of the Institute of Polymer Science and a professor of polymer science. Here he is examining the sample area of a Scanning Tunneling Microscope, which was able to display individual atoms. The atoms are visible on the screen behind him. The ovals on the background monitor screen are graphite atoms enlarged more than ten million times. *Polymer* (Fall 1990).

his wife, Bowles realized that perhaps Kelley was serious and immediately called him. Kelley said, "We are launching a major construction on a dramatic building for polymers and if anybody knows that sort of business on campus you do." Bowles protested a bit, saying he did not know anything about polymers. "You can learn 80 percent of the important concepts in polymer science in about an hour," Kelley told him. The additional 20 percent required about twenty years of research in a laboratory. They decided that night that, if Muse would allow it, Bowles would become the first associate dean of the college. Since the polymer program was so important to the university, Muse agreed. At his new job, Bowles's ability to bridge the divide between faculty and administration was a key factor in bringing the new college together. Kelley said, "The faculty just took to him . . . because he was oriented to making sure that they were served and they weren't used to that from any kind of management structure or campus services."[71]

There was an increasing number of faculty to keep happy. At the time of its completion, the new college had twenty-seven faculty members in both polymer science and polymer engineering. Kelley had long emphasized the importance of adding new faculty and increasing their funding opportunities. He said, "The importance of adding quality faculty can never be overemphasized."[72] The Eminent Scholars grant was one way the Institute of Polymer Science, and, later, the new college were able to attract world-class scholars. These were competitive grants offered through Ohio's Selective Excellence Program, the goal of which was to provide economic incentives to improve leading academic programs in the state's public universities. Governor Richard Celeste devised the Selective Excellence Program, but it was State Senator Oliver Ocasek who came up with the idea of Eminent Scholars in 1979.[73] These improvements or "enhancements" would lead, it was hoped, to increased investment by industry.

The University of Akron's first Eminent Scholar grant, in May 1984, established the school's first million-dollar endowed chair, the Alex Schulman Chair in Polymer Science. The endowment was comprised of a $500,000 grant from the Eminent Scholars Award from the Ohio Board of Regents, a $250,000 G.A.R. Foundation grant, and $250,000 from A. Schulman and the Schulman family.[74] Wayne L. Mattice, formerly a Louisiana State University professor of chemistry, held the chair first, in 1985. Mattice said, "It is important for you to understand why I came here. . . . I'm here because of the resources that are available. I've been in polymer research for twenty years and one of the best places to practice polymer research is in Ohio."[75] At Akron, Mattice directed his research toward examining the physical properties of macromolecules.

The university received a second Eminent Scholar grant when the R.C. and Katherine Musson Charitable Foundation established the Robert C. Musson Eminent Scholar Chair in Polymer Science. After a lengthy international search, the position was filled by Ronald K. Eby, who had been a professor of materials science and engineering at Johns Hopkins University since 1984, after working for the National Bureau of Standards for twenty years. His research interests included physical acoustics, advanced fibers, and computer modeling.[76] Mark Foster, current chair of the Department of Polymer Science, said Eminent Scholar grants were a "major influence on the department."[77]

A third key chair was the Harold A. Morton Chair in Polymer Physics and Polymer Engineering, established in 1987. This was the result of a gift of $1 million from Mildred Morton (no relation to Maurice Morton) to honor her late husband, a chemist and entrepreneur who owned two Akron polymer manufacturing companies in the 1940s. The endowment supported a distinguished visiting professorship as well as a permanent endowed faculty chair. Alan Gent, a principle researcher for several government agencies, including NASA, originally held this chair. He was a panel member of the National Research Council that worked on designing new solid rocket motors after the *Challenger* accident.[78]

In 1989, Kumho of South Korea established the first endowment for the Akron polymer program from a country other than the United States. At the time, Kumho was the fifteenth largest tire-making company in the world, and it pledged a $500,000 endowment, to be given in annual installments of $50,000 for ten years. Since the total college endowment at that time was $2.55 million, this was a significant gift. Several of Kumho's employees had graduated from the polymer program at the University of Akron and the company wanted to strengthen that relationship. Kelley planned to allocate earnings from the endowment to faculty development and student recruitment. Roderick Quirk is the current Kumho Professor and Distinguished Professor of Polymer Science.[79]

The polymer science faculty was involved in a number of important research projects during the 1980s. One of the most significant was work for NASA through the Battelle Memorial Institute, in Columbus, Ohio.

In 1985, NASA awarded Battelle a million-dollar-a-year grant for a projected period of five years for commercial development in space. This was a nationwide program with ten other industrial and six academic partners. Battelle researched metal alloys, glass ceramics, electronic materials, and polymers. The University of Akron assisted its researchers because so many polymer materials were to be used for the then-planned American space station. Richard Nathan, one of the Battelle directors, said, "We consider this a high priority program because space manufacturing activities will become important to the nation in the next century."[80]

Although this work was cut short after the *Challenger's* tragic explosion in 1986, the University of Akron continued to work with NASA. The university conducted a study on the O-rings that caused the failure of the space shuttle, which killed Akron astronaut Judith A. Resnik and her colleagues. Frank Kelley was a consultant to the solid rocket booster manufacturer Morton-Thiokol, and his research, funded by the company, provided key data early in the investigation. Later Alan Gent was a member of the space shuttle redesign review committee. It verified that the rubber O-rings used as seals in the solid rocket boosters failed due to cold temperatures in Florida at the time of liftoff. Hot gasses leaked through the seals, one of the primary reasons for the catastrophic explosion. The review committee made recommendations for redesigning the joint between the booster segments to avoid this critical flaw in future launches.[81]

While all of the faculty in the new college were involved in important research, they faced a significant problem that could have potentially curtailed future growth. Without a larger facility, the college would stagnate. Fortunately, just six weeks after the establishment of the College of Polymer Science and Polymer Engineering, the university broke ground on a major new building, ultimately named the Goodyear Polymer Center. The pace of change and growth at the university was a major news story in Akron and throughout northeast Ohio. Barbara N. Zimmerman, assistant to the dean, wrote, "These events—the establishment of the College of Polymer Science and Polymer Engineering and the construction of the new polymer research building on the University of Akron campus—have contributed much toward instilling Akron's citizens with a new spirit of confidence in the city's economic and industrial future."[82] That future would change Akron's skyline.

BUILDING A BEACON

Both Frank Kelley and William Muse knew that if their master plan was to become a success, the campus required significant facilities improvement. Remaining at the leading edge of polymer research in the United States required advanced laboratory and classroom spaces.[83] But, by the mid-1980s, the Auburn Science and Engineering Center and Whitby Hall, which were used by other academic programs at the university as well, could no longer meet the needs of the polymer program. Kelley recalled that there was a real "space crunch."[84] The university needed a new building, so officials decided not only to improve existing facilities but to make a bold architectural statement that would become a symbol of excellence. Muse wanted it to become a "beacon of hope for the region."[85] Kelley thought it would be an "icon to the importance of this region to polymer science."[86]

The efforts began with the establishment of a capital plan and its approval by the state's Board of Regents. The polymer building was the main item on the list of capital requests, and the plan was eventually approved. While the state initially planned to give only $5 million, it later increased that amount to $15 million for the project, with $2.6 million coming from university fund-raising. The other early major decision was where to locate the building. The first proposal was to locate it in the Olson warehouse, which the university had just purchased, on the north side of campus. Preliminary planning joined the proposed building as an annex to the Olson facility, which housed polymer engineering. Muse eventually rejected this plan; he did not want to hide the new facility. He said, "I wanted this building to be a visible symbol of the renaissance of Akron."[87]

At the same time, there was another debate on the future of Buchtel Avenue. Muse believed Buchtel Avenue had been a problem for the university for many years. Its four lanes ran right through the heart of campus, with the main library and dormitories on one side, and the academic and administrative facilities on the other. It was a very congested road made all the more dangerous with students constantly trying to cross it. One of Muse's first goals upon his arrival on campus was to close the road. This would do two things for the university: it would remove a potentially dangerous road and eliminate the possibility of a student being needlessly injured or killed, and, second, it would create a chance to beautify the urban campus by replacing the road with a landscaped pedestrian walkway. But with the closure came another unexpected opportunity.

After Muse rejected the first proposed location for the polymer science building, Donald Bowles took a large map of the campus home and spread it out on his dining room table. The effort to close Buchtel Avenue was taking place at roughly the same time as the planning for the polymer science building. Bowles and his staff realized that when the pedestrian walkway was completed there would be a triangle of land on the west end of campus, where Buchtel, Center, and Sumner Streets came together. Muse recalled that sometime later Bowles returned to him with a new idea, suggesting "Why don't we put it right in the pathway of Buchtel Avenue, where it can be clearly visible not only from downtown and from the campus but from miles away." [88] Muse agreed immediately. From this vantage point, the new polymer building would overlook the city and become a striking visual symbol.[89] In November 1984, Bowles asked for and received permission from the university's Board of Trustees to place the building in the new location.[90]

In September 1985, the university officially entered the design stage for the new polymer building.[91] Richard Fleischman Architects, of Cleveland, won the bid. Muse was very clear that he did not want another rectangular, yellow, brick building. He told Fleischman, "We're looking into the future here. We're looking at the future of not only Akron but of northeastern Ohio and I want something that stands out."[92] Fleischman identified his own style as "very modern." He often used simple geometric forms to define his structures and built a steel skeleton covered by reflective and transparent glass.[93] Fleischman said, "Everything I do has to have a dream. Every architect takes vision into reality. . . . It's a symphony of space—we as architects are

Artist's rendering of the polymer science building, with E.J. Thomas Performing Arts Hall to its near left and the National Inventors Hall of Fame to the far left (circa 1988). *College of Polymer Science and Polymer Engineering Archives.*

conductors of spaces. When I create space, I think, 'How do I want people to feel?'"[94] Kelley contributed his vision as well. Over dinner one night, Fleischman sketched out his two-towered plan, essentially two squares with a connecting bridge in the middle. Kelley expressed his own belief that the structure needed more curves. Fleischman began modifying his sketch and the two came up with the curved features that would eventually become the footprint for the atria.[95] Fleischman believed this give and take with the client was essential to transforming his architectural vision into a reality that would work for those who would occupy it.

The practical goal for the new building was to accommodate effectively the changes and growth of the polymer program at the university.[96] But Fleischman wanted the people inside the building to be inspired to investigate new areas of science and engineering. The building's open spaces would cultivate their creativity, he hoped, and foster the sharing of ideas. Its openness mirrored the increasingly interdisciplinary nature of the polymer program itself. The new building's design

Architect's model of the polymer science building (circa 1988). *College of Polymer Science and Polymer Engineering Archives.*

attempted to bridge the divides between academic specializations, with a unifying unpartitioned layout. Of the more than three hundred structures he built in his forty-year career, Fleischman selected the polymer building for the cover of his own retrospective book documenting his life's work. The title of the book could have also been the inspiration behind his university project—"Spaces to Be Shared."[97]

Very early on, its planners dismissed one unique idea for the building.[98] Fleischman said, "Originally, we wanted to do the entire building in polymers."[99] Since the building's function was to be used for the polymer program, many people thought that polymer materials should be used to construct it. The problem was that the use of polymer materials as a structural element in such a large building was untested.[100] To ensure the safety and durability of these materials, would have added another two years to the construction time of the building and cost an additional $7 to $10 million. Kelley was disappointed but agreed with the decision to construct the building out of traditional steel and glass. He said, "It's a twenty-first century building, but unfortunately we can't use twenty-first century building materials."[101] The much publicized "plastic building" was not to be.

The disappointment over the decision to use traditional building materials faded by the time actual construction began. Buchtel Avenue officially closed in November 1987, and ground was broken on April 15, 1988, with Muse presiding over the occasion. Muse told the gathering of politicians, university leaders, community members, and students that the new building would propel the university and the city to greater heights. "Completion of the building," he said, "will enhance the research environment for our polymer faculty and enable us to contribute more significantly than ever to the rapidly growing polymer industry."[102]

The building quickly took shape. In January 1989, the steel structure was topped off, a high point in this process. On a cold windy day, workers hoisted the final thirty-foot steel beam to the top of the 190-foot structure, a small Christmas tree on one end and an American flag on the other. After two and a half months, the ironworkers finished their part of the job in front of the assembled news media and university officials. Kelley, who had been born on that same day in January, called it "an excellent birthday present."[103]

Construction continued for almost two years after the topping-off ceremony. It was at such a visible point on the Akron skyline that the entire city was aware of its progress. The building had connected towers, one twelve stories tall and the other nine stories, with 146,000 square feet of usable space. This space was distributed through fifty research laboratories, twenty faculty offices, ten faculty laboratories, and

President William V. Muse and Dean Frank Kelley at the groundbreaking of the polymer science building in 1988. *College of Polymer Science and Polymer Engineering Archives.*

The polymer science building under construction (circa 1988). *College of Polymer Science and Polymer Engineering Archives.*

twenty-five additional offices that could accommodate two hundred students. It had a 213-seat auditorium and an atrium on each floor. One of its most distinguishable features was a slanting canopy on the top of the building, used to protect the mechanical equipment on the roof and to tie the structure together visually. Fleischman came up with the slanting architectural element after studying the adjacent E.J. Thomas Performing Arts Hall, which had been the most architecturally significant building on campus, with its long, graceful, sloping roof. By imitating those angles, Fleischman found a way for the buildings to relate to each other (a symbolic bridge between the arts and the sciences), while at the same time maintaining their distinctive aesthetics.

In total, the building cost $17 million.[104] It became the fourth tallest building in Akron, behind the (then named) twenty-eight-story First National Tower, the twenty-four-story Cascade building, and the nineteen-story Ohio Edison Building.[105] Most of the funding came from the state of Ohio, which provided $15 million. The remaining $2 million came from sixty-seven private sources, with fund-raising assistance from Robert

The dedication of the polymer science building took place on April 1, 1991. Shown here, from left to right, are architect Richard F. Fleischman, President William V. Muse, Dean Frank Kelley, and Associate Dean Donald L. Bowles. *Polymer* (Fall 1991).

Mercer, the former chief executive officer at Goodyear. Some of these private gifts led to named areas of the building that represented the major corporations in (and some outside) the rubber industry. These included the Goodyear Auditorium, the Bridgestone/Firestone Lobby, the GenCorp Lobby, the B.F. Goodrich Lounge, the Columbian Chemicals Lounge, the AT&T Classroom of the Future, and the Uniroyal Goodrich Tire Company Student Lounge. The new building housed a polymer program that had doubled in size over the previous decade. It had twenty-eight faculty and over 250 postgraduate students. The entire program received $7 million in funding the year the building was dedicated. This included $2.5 million from gifts and $4.5 million for funded research.[106] This was nearly a sevenfold increase from 1978 when Kelley took over the Institute of Polymer Science.

The responses to the new building from those both in and outside the university community were very positive. The *Chicago Tribune* re-

The Goodyear Polymer Center, reflecting clouds and the university buildings to its east. The Donald L. Bowles Memorial Garden is at the forefront. *Photo by Mark D. Bowles.*

ported that it seemed like "a vision of the Akron of the future."[107] One Associated Press writer thought that, with this new building, the city "pinned many of its hopes to the polymer industry."[108] It represented a way for Akron to become distinctive again. An AT&T representative said the building added a "majestic sparkle to the Akron skyline."[109]

Internal praise at the university was just as high. Kelley said, "If anything, it surpassed all of our expectations. It's sort of a dream come true. How often do you get to help design the ultimate workplace?"[110] Muse was equally enthusiastic. "It's a significant step for us in solidifying our position as the University in the field of polymer science," he said. "The whole facility will pretty much put us in a class by ourselves in terms of environment."[111]

State and national politicians also firmly believed in the building's importance as a leadership center for the development of polymer science research. Governor George Voinovich said, "It is an investment that will pay handsome dividends to the State of Ohio—in terms of employment, research, and company expansions—well into the next century."[112] Congressman Tom Sawyer believed the building would reestablish "the importance of the university at the heart of our economic development."[113]

After the new Goodyear Polymer Center was completed, the university began proclaiming that "This tower is not an ivory one." The message reinforced the goals of Polymer Valley. The new building was to be a "symbol of the bond between the University of Akron and Ohio citizens."[114] As a focal point on the local horizon, the building stood as a beacon at the summit of Polymer Valley.

CONCLUSION—POLYMER VALLEY IS HERE

William Muse left the University of Akron to become president of Auburn University just one year after the Goodyear Polymer Center was finished. Muse said, "I felt certainly by the time I left Akron, we had made tremendous progress." Polymer Valley was becoming a reality. Upon his departure, the *Akron Beacon Journal* debated his legacy. To some, he was considered a "visionary," the person who "pulled the University of Akron out of its small, parochial past and into a promising future."[115] Even his detractors conceded that Muse attracted more national attention to the university than any of his predecessors.

During Muse's tenure, and certainly since Kelley's return in 1978, some of the polymer program's major goals had been achieved. The Institute of Polymer Science was thriving in its new building. The

The Goodyear Polymer Center at night. Frank Kelley called it a "visual beacon for the city and a symbolic beacon for the region." *Polymer* (Fall 1991).

EPIC consortium was growing and making strides by the day. The new polymer science chairs and endowments were attracting world-class faculty to the university. The polymer faculty had won twelve national and international medals for distinguished work in their fields. A first-ever College of Polymer Science and Polymer Engineering had been established, and its graduates were taking prestigious leadership positions in organizations such as DuPont, Lawrence Livermore Laboratory, the California Institute of Technology, Wright Patterson Air Force Base,

Allied Chemical Corporation, Dow Chemical, and Union Carbide. No other institution in the country had graduated more polymer Ph.D.'s than the University of Akron.[116]

But, what did the polymer program mean for the struggling Akron economy which had suffered so much since the loss of tire manufacturing in the city in 1978? At an Akron Roundtable meeting in the early 1990s, Frank Kelley addressed this question exactly. Looking back over the previous decade, he said, the University of Akron had achieved significant milestones. However, he asked, what have these milestones "really meant in the context of a city, reeling from the loss of its manufacturing base?" He answered his question by stating that Ohio had recently become the number one state in the nation in the plastics field. This, he believed, was emblematic of the increasing economic strength of the region. Over one hundred thousand people now worked in the plastics industry in Ohio, resulting in the shipment of $16 billion of goods per year. Plastics companies came or remained in Ohio for several reasons: the academic excellence of the university, proximity to other plastics companies, a skilled labor supply, and regional assets the industry needed to thrive. Numerous company officials in the polymer industry had told Kelley that "their decision to stay in this area has been positively influenced by the polymer focus [at the university]."[117]

Furthermore, there were some positive developments in the rubber industry itself. Unlike the end of the 1970s, the end of the 1980s represented a surprisingly optimistic period. From 1978 to 1988, the major rubber companies cut 61 percent of jobs in Akron, a drop from 28,450 to 11,000. But in 1989, the tide appeared to be turning. Goodyear had just survived a hostile takeover attempt by James Goldsmith. Its world headquarters would remain in Akron. Based on the successes at the University of Akron, the thriving plastics industry in Ohio, and the commitment of Goodyear to the city, Kelley concluded, "We have truly become the Polymer Valley."[118] Could this momentum continue through a new president at the university and another decade of change?

6

APPLYING POLYMERS

1983–2001

IN 1942, AN ARTICLE ENTITLED "PLASTICS COME OF AGE," appeared in *Harper's* magazine.[1] Its writers made a bold argument that in the near future synthetic plastic substances constructed in laboratories would have a revolutionary impact upon everyone's lives. They prophesied that "plastics may one day become a dominant material" and even possibly become as "dominant as metals." When scientists mastered synthetic materials, plastics would have "come of age." Those predictions have become reality, but what enabled the Polymer Age to blossom was something more than science. Scientists were responsible for identifying the structure of macromolecules, characterizing their composition, producing small amounts of revolutionary new polymer materials, and perfecting how to manipulate them. But it was necessary to go beyond what science alone could achieve. What was required was the production of polymers on a large scale, the development of useful new products, and the consumption of them by a mass audience. This is the domain of polymer science's sister discipline, polymer engineering.

The importance of polymer engineering cannot be overestimated. Frank Kelley said, "Polymer science really means very little unless it's tied to its usefulness in society."[2] The ability to apply polymers is what elevates the esoteric basic science of macromolecules into a practical and useful element in our everyday lives. Without polymer engineers, this would not be possible. Eric Baer, a polymer scientist from Case Western Reserve University, has emphasized that polymer research is itself an "applied field." Its primary indicator of success is that it "takes the science of today and makes it into the engineering and the technology of tomorrow."[3] Polymer engineering accomplishes this through various types of laboratory processing techniques, including compounding and blending. Processing can take place with massive machines such as injection molders and extruders to create complex and sometimes very large products. It also exists on the very small molecular (or nano) scale where resulting compounds can be combined in a variety of polymer applications. Polymer engineers believe they are working in a critically important era because they are finally having success integrating and developing polymer materials across the entire size scale.

Despite this current success, it was necessary to overcome one essential historical problem with polymer engineering. Sadhan C. Jana, chair of the Department of Polymer Engineering at the University of Akron, said that for many years "Polymer engineering as a discipline did not exist." Physicists, along with mechanical and chemical engineers, often explored specific polymer problems. But there was no dedicated academic program in polymer engineering that matched the sophistication of polymer science. In the early 1980s, the University of Akron decided to direct more attention to establishing and cultivating its own expertise in this still emerging field. As a result, Jana said, "The landscape for polymer engineering has changed completely."[4]

In *What Engineers Know and How They Know It,* author Walter Vincenti explains, "Engineering knowledge, though pursued at great effort and expense in schools of engineering, receives little attention from scholars in other disciplines." When the engineers' work is analyzed, it is often shaped by the bias that they are simply handmaidens to their scientific brethren and are merely applying knowledge bestowed upon them to "fashion material artifacts." Vincenti and others have argued for a new understanding of engineering knowledge as its own "significant component of thought."[5] Engineering is not the routine application of knowledge handed down by science, but instead represents unique, creative, and constructive thought in its own right. Although engineering may "*apply* science," Vincenti concludes, it "is not the same as or entirely *applied* science."[6]

By "applying polymers," engineers at the University of Akron promoted the growth of a new discipline, cultivated advanced faculty research, provided doctoral-level training and research opportunities for its students, and ultimately contributed to the economic vitality of a region. They were not merely following the lead of polymer scientists but were carving out their own distinct discipline.

COMMERCIALIZING RUBBER AND POLYMERS

The late 1830s marked a turning point in the cultural obsession with rubber. While rubber's amazing qualities had been well known for centuries, most commercial products made with it were critically flawed until the secrets of vulcanization were understood. Nevertheless, Americans still invested heavily in rubber products that cracked in the cold and melted in the heat. Author Charles Slack noted, "The American obsession with rubber during the 1830s bears a striking resemblance to the Internet frenzy of the 1990s."[7] During both of these eras, investors were willing to risk their fortunes on still unproven technological ideas. It was not until 1839, when American inventor Charles Goodyear and British researcher Thomas Hancock independently established the vulcanization process (combining rubber with heat and sulfur to cure it), that rubber products actually became useful and reliable. According to historian Page Smith, the result of this discovery was an incredible commercialization of rubber, including "an abundance of rubber shoes, raincoats, hats, rubber gloves, pails and drinking cups, tents, water beds, life preservers, inflatable rubber rafts, hot water bottles, teething rings, 'cushions of every description,' nipples for babies' bottles, elastic bands, 'errasive rubber,' syringes, and washers."[8]

After 1839, there was a revolution in the use of *natural* rubber for commercial products, but an understanding of the importance of *synthetic* materials came more slowly. In the nineteenth century, classical chemists prized natural materials that could be crystallized, as opposed to unknown substances that could not be so identified. For example, a chemist in Robert Bunsen's laboratory in the 1840s described the "beautiful crystallizable" compounds he discovered after certain reactions.[9] Aesthetics aside, chemists depended analytically upon these types of crystals for purification purposes.[10] They represented the perfection of nature, and chemists believed they were learning how to become masters of them. But all too often they found something else after their chemical reactions that became a bane of their laboratories—gums, syrups, and resins. Chemists could neither purify nor characterize these substances, and organic analysis appeared useless.[11] These residues were ugly, formless goos that stuck to the base of the chemists' pots and adhered to the sides of their test tubes. When they tried to get rid of the goo by washing it away, it clogged their laboratory drains.

By the twentieth century, researchers began to realize what these substances actually were. Instead of formless masses, they were actually a series of long intertwining chains of chemical units. It was not until the 1920s that chemists began to understand that these substances were "macromolecules" or "polymers." With this knowledge, it soon became apparent that an entirely new realm of material mastery was possible. This resulted in a revolutionary opportunity to manufacture new types of products, and polymers have today become a dominant substance in our lives. From laboratory bane to material dominance, the history of the attempt to transform natural and synthetic rubber into useful products is the compelling story of the emergence of polymer engineering.

Although the discipline known as polymer engineering was not established until the latter half of the twentieth century, one of the first efforts to engineer and apply polymers actually began in 1863 with John Wesley Hyatt, a printer in New York. Hyatt heard that a law firm was offering a prize of $10,000 to anyone who could develop a material that could substitute for ivory in billiard balls. He immediately thought of a substance called "collodion," discovered in 1846, which was the first synthetic material made from cellulose nitrate. Collodion revolutionized the photographic industry, and by using it to make emulsions on glass plates, it replaced the daguerreotype process. Hyatt, recalling that he had once spilled some of this collodion and it had become a hardened mass, tried to transform collodion into a billiard ball to win the prize. He experimented with a number of different mixtures, eventually mixing camphor with nitrocellulouse, heating it, and compressing it into a mold. Hyatt had his billiard ball, and the world had its first polymer product. He called the material "celluloid."

Hyatt later invented a "stuffing machine," a forerunner of injection molding, to create a wide variety of celluloid objects. He soon branched out from billiard balls to other products, including combs, handles for knives and brushes, piano keys, harness fittings, and any other novelty previously made of ivory, tortoiseshell, or horn. The new material was inexpensive to produce and made highly decorative items available to the masses for the first time. Cultural historian Jeffrey L. Meikle recently wrote, "Awareness of substitution, imitation, and innovation as three distinct motives for using plastic emerged almost simultaneously with Hyatt's invention of celluloid."[12]

These early products were the result of modifying a natural polymer like natural rubber or cellulose. Not until 1907 did Leo Baekeland, a chemist, create the first synthetic polymer with no direct natural counterpart. He achieved this through his research into the condensation reaction of coal tar (phenol) and a derivation of wood alcohol (formaldehyde). Although others had experimented with this reaction, his new "heat and pressure" approach yielded a revolutionary new product. He called it "Bakelite," and it was immediately used in many applications, most notably as an insulator for electric connections.[13] By the early 1920s, his Bakelite Corporation, whose corporate logo was the infinity symbol, had issued a pamphlet describing the various—nearly infinite—ways its product entered the lives of the typical family. The list included telephones, radio parts, teething rings, shaving brushes, pens, and gearshift knobs in automobiles. The extent of Baekeland's influence and significance is hard to overestimate. What he achieved, according to historian Robert Friedel, was "so very different from anyone before him that the very stuff of which the world was made began to change."[14] *Time* magazine put Baekeland on its cover in 1924 and fifteen years later named him the "father of plastics."[15] Efforts of Hyatt and Baekeland began the trend of replacing natural materials with polymer substitutes and led to what eventually became polymer engineering.

The machinery necessary to manufacture these commercial products—processing and shaping machines—was also essential to the foundations of a new discipline. Most of the basic machinery necessary to process rubber had been invented in the nineteenth century. Thomas Hancock, the British vulcanization co-discoverer, developed the first essential piece of rubber manufacturing equipment. Called Hancock's pickle, it converted rubber scrap into a reusable doughlike material. After a procedure known as "mastication," the "pickled" rubber became an important component in processing high-quality rubber. Initially, the pickle was manually operated. Hancock solved this problem in 1840 by inventing a power-driven single-roll masticator.

The pickle was similar to the rubber "calendar" developed in America in the 1830s by Edwin Chaffee. The calendar had several rollers which could produce very thin sheets of rubber. Chaffee also invented a two-roll mill in which the rubber was masticated through a shearing action as it passed between two rollers moving in opposite directions at different speeds. One major advance solved the problem of the masticated rubber's exposure to air. In 1916, Fernley Banbury developed a replacement to the open mill, inventing the Banbury internal mixer. According to John Loadman, "The calendar and two-roll mill remain the basic equipment of virtually every rubber-manufacturing facility today. . . . the fundamentals remain the same now as they were one hundred and fifty years ago."[16] Likewise, the Banbury still remains the most used mixer for rubber compounding.

After the rubber had been processed, the next step was to press it into a shape required of a commercial product. In the 1850s and 1860s, several types of presses were developed in which rubber was shaped in a mold and then vulcanized. In compression molding, the rubber was placed below a mold that descended on the rubber so that it filled the empty space. This type of molding was a batch process, but it could not produce products with long shapes, such as a hose or a seal. To achieve these results, another type of machine known as an "extruder" was required. In 1881, Francis Shaw invented his screw extruder in which a piston forced a piece of compounded rubber through a die. The die determined the final shape of the product as the rubber squeezed through the die and cured in air ovens. It was not until the 1940s that these two types of machines—presses and extruders—were combined into what is known as injection molding. These machines would form the basis of a new discipline.

The study of polymer engineering did not emerge at the same time these tools were invented. Instead, it evolved as a hybrid of other intellectual disciplines and eventually established itself as a specialization of its own toward the latter part of the twentieth century. Whereas in the United States polymer science traced its roots from the broader discipline of chemistry, polymer engineering descended from the study of chemical engineering. In Germany, polymer engineering was closely associated with mechanical engineering, and there were important ties in the United States to this field as well. Polymer engineering was also related to metallurgy in some respects. Many polymer engineers used the same experimental tools, such as x-ray diffraction and electron microscopes, as did metallurgists.[17] Polymer engineering at the University of Akron, like the broader field, evolved from chemical engineering and used the processing, molding, and shaping machines developed in the nineteenth and early twentieth centuries to work toward the establishment of a new, independent discipline.

THE EMERGENCE OF POLYMER ENGINEERING IN AKRON

While the natural sciences were taught at Buchtel College almost from its founding, engineering lagged by several decades. The first engineering courses at the University of Akron were not offered until 1914, when the school established its College of Engineering—exactly forty years after the first science course was taught. Parke Kolbe, the university president, considered this new college one of his proudest achievements. One reason for his enthusiasm was the new cooperative educational relationship between the engineering college and local industries. The rubber companies were the first to take advantage of the program. Goodyear, Firestone, Goodrich, Akron Rubber Mold, and Miller Rubber, among others, established thirty industrial scholarships for a cooperative work plan. These were available to qualified graduating high school students.[18] The company paid the student's tuition, and every two weeks he worked in different factory production departments. The students were selected in pairs so that together they would essentially hold one full-time position in the rubber factory.[19] The student was paid $75 per month, and completing the program took nearly five years. The University of Akron's 1917 annual report stated: "The establishment of these scholarships seems . . . to point toward what will be considered by most people as the Utopian dream of a visionary college professor."[20]

Soon the first students graduated from the university's new engineering departments. John S. Kennedy, a civil engineer who graduated in 1918, was the first to earn his degree from the College of Engineering. The first mechanical engineer received his diploma in 1919, along with the first graduate in manufacturing production (industrial engineering). In 1917, the university established the Department of Electrical Engineering, and James E. Lynn was its first graduate, in 1922.[21]

The engineering program had remarkably stable leadership, as over the next fifty years only two men served as dean of the College of Engineering—Frederic E. Ayer (1914–1946) and Ransom D. Landon (1946–1964). Major changes began under the college's third dean, Michael Rzasa, starting in 1964. Rzasa had worked for seventeen years in the petroleum industry, and he decided it was important to make some structural changes in the program. He began by laying the groundwork for a department of chemical engineering, an important first step toward a polymer engineering program. A second key Rzasa initiative was the graduate school. The master's program was only eight years old when he arrived, and Rzasa wanted not only to increase the number of these students, but also to establish a Ph.D. program.

The doctoral program commenced in 1967, and Rzasa awarded the first graduates their degrees in 1970. One year later, the College of Engineering moved into the new Auburn Science and Engineering Center, where it shared space with the Institute of Polymer Science. This created a greater opportunity for practitioners of the two disciplines to interact, especially the chemical engineers and the polymer scientists.

During this same period, the University of Akron began actively to promote research in polymer engineering. In 1964, the polymer chemistry program became an interdisciplinary program in polymer science that offered three main areas of study. Students working toward a Ph.D. in polymer science could select from a chemistry, physics, or engineering program, which enabled engineering graduates to enter the polymer Ph.D. program while maintaining their area of specialization. The engineering classes included polymer technology, polymer processing, design of rubber components, mechanics, and advanced engineering materials.[22] In August 1964, the University of Akron announced in *Chemical and Engineering News* that these new engineering opportunities were available for students.[23]

In 1966, the University of Akron's polymer program began another engineering initiative with its Applied Research Section. In that year, the U.S. Department of Agriculture awarded Howard Stephens, manager of applied research, a three-and-a-half year, $178,750 grant to explore the possibility of using starch and its derivatives to reinforce synthetic and natural rubber. While carbon black had been used as the primary reinforcing agent for rubber since the early part of the century, Stephens planned to determine if the starch provided any benefits.[24] His study became the first project taken over by the newly created Applied Research Section of the Institute of Polymer Science. All previous research consisted of basic or fundamental research; his applied research was the forerunner of the polymer engineering program.[25]

Alan Gent, a professor of polymer physics at the University of Akron, became an early and active promoter of polymer engineering. In 1967, the American Society of Engineering Education asked Gent to speak on the general subject of "Engineering with Plastics." Gent eagerly accepted, writing to Dean Rzasa, "This seems like an excellent opportunity to publicize the University of Akron, our Engineering College, and our Polymer Science and Engineering Program."[26] At the time, the program was very informal, consisting of graduate level courses for mechanical and chemical engineers. The establishment of the interdisciplinary program in polymer science further emphasized aspects of engineering, and Gent continued to promote these activities to the broader engineering community. In 1967, he wrote a letter to the chairman of the Materials and Processing Committee of the American Society of Electrical Engineers, emphasizing how important this new field was at the University of Akron. "We are thus committed to a program of education in polymer engineering," he wrote.[27] However, Gent received very little academic support at the time.

When Frank Kelley arrived at the university in 1978, he brought a renewed interest in expanding the role of polymer engineering, believing it had played handmaiden to polymer science for too long. In one of Kelley's first departmental reports, entitled "Plan for the '80s," he complained that the emphasis on the science of polymers minimized the engineering of polymers.[28] Despite the efforts of those involved in the interdisciplinary program in polymer science to bring engineering to the forefront during the 1960s, Kelley and others knew that success had eluded them. Carl Snyder, vice president of research for Goodyear, said in the early 1980s that the University of Akron was excellent in polymer chemistry, among the top five institutions in the world. But, he said, the program had a significant problem: "In polymer applications, they have been weak."[29] Only a few professors on campus at the time were actively engaged in polymer engineering research. (Eberhard Meinecke and Alan Gent were notable exceptions.)

The faculty shared these assessments and said that, while their achievements were "notable in polymer science," the graduate training of engineering students was by contrast "more limited" in terms of sheer numbers and also in the depth of training. A new emphasis upon engineering would benefit the overall university polymer program in two ways. First, it would broaden the program's academic scope and offer more opportunities for students and faculty research. Second, based upon the future funding trends, more research dollars would come in from industry. Polymer companies often experienced problems related to polymer processing and composite materials, problems an academic program could help solve. Therefore, a stronger polymer engineering program could potentially expand industrial funding sources for the university. The leaders of the polymer program decided it was time to devote more attention to engineering and develop an administrative structure in which it could grow and thrive. They needed a dynamic individual on campus who would take ownership of the polymer engineering program and put it on the path to success. That individual was James White. According to Sadhan Jana, White was a "visionary" who established a new polymer engineering laboratory on campus. "Personalities like this," said Jana, "are born only once in a while."[30]

James White was born on January 3, 1938, in Brooklyn, New York, and became interested in polymers while working on his bachelor's degree in chemical engineering at the Polytechnic Institute of Brooklyn. An elective course in polymer chemistry with polymer pioneer Herman Mark during his senior year in 1959 solidified his desire to pursue an academic career in the field. After graduation, he earned master's and Ph.D. degrees at the University of Delaware, completing his studies in 1965. After working for two years in industry, he took a position at the University of Tennessee as a professor in the Department of Chemical and Metallurgical Engineering. During his sixteen-year tenure there, his long-term goal was to establish a polymer engineering program. By 1976, he was granting Ph.D.'s with a polymer engineering concentration through the chemical engineering department. By 1983, he had thirty-five students in his program.[31]

At this point, White became dissatisfied at Tennessee. He wanted the university to commit more strongly to his program, but instead felt it had more interest in athletics than polymers. Kelley knew of his dissatisfaction and invited White to come to Akron to teach a course in chemical engineering, and, potentially, to stay to head up its limited polymer engineering program. The local rubber industry also welcomed him; the research heads of B.F. Goodrich, Goodyear, General Tire, and Firestone held a dinner organized by Frank Kelley to convince White to relocate to Akron. White, in his trademark bow tie (a regular accessory for twenty-five years because he discovered traditional ties were easily caught in rubber machinery), eagerly accepted the opportunities offered by a move to Akron.

James L. White was appointed head of the polymer engineering program in 1983. He has served in many leadership capacities including department chair (1983-1997), director of the Institute of Polymer Engineering (1983-2001), and Harold A. Morton Professor of Polymer Engineering (1997-present). Here he examines a new Leistritz counter-rotating extruder. The machine, located at the Sidney L. Olson Research Center, was acquired though a $65,454 donation from Leistritz and $80,000 from the college's equipment budget. *Polymer* (Fall 1991).

In July 1983, White established the new Polymer Engineering Center at the University of Akron, bringing six students with him from the University of Tennessee. "At the end of 1982, polymer engineering had one-fourth of a person," White recalled. "There were twenty-five at the end of [1984]."[32] He had high hopes for the program. Akron could "become the center for a renaissance of the American polymer industry—if not for the actual manufacturing plants, then certainly for the design and development of new instruments and machinery for such plants," he said. But White cautioned that thinking up new ways for a rebirth in Akron was a favorite "parlor game" in the city.[33] He worked tirelessly to ensure that the shell of the polymer engineering program he created in 1983 would become a solid foundation from which the discipline could grow.

BUILDING THE POLYMER ENGINEERING PROGRAM

The initial funding for the new Polymer Engineering Center came in part from eight rubber companies. In December 1983, these organizations pledged a total of $1 million in support over the coming three years: Goodyear ($300,000 in funds and $400,000 in equipment); Firestone ($100,000 in funds); B.F. Goodrich ($75,000 in funds); Monsanto ($75,000 in equipment); General Tire ($60,000 in funds); Gates Rubber ($15,000 in funds); and Armstrong Rubber ($3,000 in funds). Louis A. Hill, Dean of the College of Engineering, was pleased with the level of corporate giving and said it was "pretty close to what we had hoped for."[34] One of White's major goals for these gifts, and for the immediate future, he said, was to "improve the design of existing process machinery and to develop new technologies."[35]

But there was a more pressing concern. If the University of Akron was to develop polymer engineering program, it would need to provide it with physical facilities. Hill and White began exploring the campus together to find a suitable location.[36] Tucked away behind the E.J. Thomas Performing Arts Hall was the old Olson Electronics Warehouse. Originally built in 1928 as a produce warehouse, the two men thought it had great potential as a home for polymer engineering. The university had acquired the building in 1979 and because of its size, over 72,000 square feet, used it for diverse activities, including an indoor practice facility for the marching band. The second floor housed art department classes, and the Kenley Players often built large sets for E.J. Thomas Performing Arts Hall productions there. But, in 1985, the University of Akron transformed the Olson Electronics Warehouse into a new space for its Polymer Engineering Center. Although the local papers reported that White was in "polymer heaven," in reality

President William Muse presents Sidney Olson with a replica of the plaque renaming the polymer engineering building the Sidney Olson Research Center. *Polymer* (Spring 1990).

he confronted very humble surroundings. On his first visit to the warehouse, White fumbled in the dark just to locate a light switch. This did not discourage his aspirations to establish what he hoped would become "the best polymer engineering program in the country."[37]

To redevelop the building and transform it into a leading research facility, Sidney L. Olson donated $1 million. The first step in the renovation process was to give the Olson Warehouse a new face. This included a new exterior shell with several inches of Styrofoam that served to provide insulation as well as a more attractive appearance. The university also accepted donated equipment such as an injection molder, a screw extruder, tubular blown extruder, and melt spinner. The first floor became the Rotational Molding Development Laboratory. Rotational molding is different from other types of molding techniques (injection molding, blow molding, and thermoforming) because it uses powdered thermoplastics and low pressures. Researchers used it to investigate "warpage" and how to eliminate it.[38] Also included on this floor were a variety of twin screw extruders and a large-scale mixer.[39] Eventually it would also house a mini-scale twin screw extruder that allowed compounding to be done on as few as ten grams of material.

Injection molding is an important process in the polymer industry. Students receive hands-on education and conduct research in the polymer engineering laboratories. Polymer engineering doctoral student Yi Zhang (left) and John Jenkins and Dave Schworm (mechanical engineering students) operate an injection molding machine on loan to the University of Akron's Institute of Polymer Engineering from the Van Dorn Plastic Machinery Company. *Polymer* (Spring 1993).

Once these were compounded, they could be taken to a mini-scale injection molder. On the mezzanine, the university constructed a two-floor model factory. Frank Kelley called this new area a "miniplant" and said that it was like a kindergarten where new polymer ideas could flourish. The factory served as a middle staging area that linked basic research coming out of university laboratories to development by local industry. The Institute of Biomedical Engineering moved into the third floor.[40] In 1990, the university completed the renovations and renamed it the Sidney L. Olson Research Center.[41]

While the Olson Research Center was being completed, the program was undergoing key structural changes as well. In 1988, it moved from the College of Engineering to a home in the College of Polymer

Extruding and blowing a bubble film in the polymer engineering laboratories (left to right, students Kaan Gunes and Rishi Kumar). *Photo by John Ashley.*

Science and Polymer Engineering. White wanted to mimic the structure of the polymer science program and create two separate polymer engineering entities—an institute and a department. As a result, he replaced the former Polymer Engineering Center with the new Department of Polymer Engineering and Institute of Polymer Engineering. These served the same functions in polymer engineering as they did in polymer science. The institute's focus centered upon research and its related activities, while the department's focus was instruction, along with recruitment, financial support, curriculum development, academic assessment, student advising, and faculty support. A chairperson served as head of the Department of Polymer Engineering, while a director guided the Institute of Polymer Engineering. Over the course of the last twenty-three years, the leadership of the Institute and Department of Polymer Engineering has been remarkably stable. Other than James White, only three other men have served in leadership positions.

Although the formation of the new college presented a tremendous opportunity to advance and strengthen the entire program, it did meet with some resistance from the College of Engineering. The new polymer college required a significant administrative reorganization: polymer science was removed from the College of Arts and Sciences, and polymer engineering from the College of Engineering. It was difficult for these already established colleges to give up academic programs, but it was most difficult for engineering. The dean of the engineering college, Louis Hill, had originally worked to recruit White from the University of Tennessee and helped him build the program from virtually nothing over a period of five years. Although he never expressed his dissatisfaction publicly, he resigned suddenly at age sixty, several months after the new college was established. University Provost Frank Marini said, "I was surprised when he wanted to take early retirement. I just didn't know he was thinking about that." Frank Kelley said, "I think it will be a devastating loss."[42]

Throughout the next decade, the College of Engineering continued to cope with the loss of its polymer engineering program. Nicholas Sylvester, who took over as dean of the College of Engineering after Hill, said, "There's jealousy. It exists. They feel the polymer science people are given preferential treatment. Some of that has happened." Irving Miller replaced Sylvester in August 1995. He too wrestled with the reality of having a renowned polymer engineering program that was not part of the College of Engineering. He said, "The engineering college is not badly thought of around the country. But if people think well of it, it's because they think that polymer engineering is a part of it."[43]

Despite the difficulties, the formation of the new college was essential for the growth of the polymer engineering program. One component of this was a strong faculty. Initially, White recruited two other founding polymer engineering faculty members (Thein Kyu and Avraam Isayev) and two founding principal engineers (Mukerrem Cakmak and Kyonsuku Min). When they arrived in 1983, they also encountered the challenges of meager facilities at the birth of the polymer engineering program. Kyu said, "In the beginning, our building had only four walls. . . . Today, we are proud we survived and are doing exceptionally well."[44] When he first came to Akron, Isayev said, "The biggest challenge was to create and build my own research group from scratch, obtaining the needed equipment—such as injection-molding machines and extruders costing thousands of dollars each—and recruiting and

(*clockwise from top left*) Thein Kyu. *Polymer* (Fall 1990). Avraam Isayev. *Polymer* (Fall 1991). Kyonsku Min-Cakmak. *Polymer* (Fall 1991). Mukerrem Cakmak. *Polymer* (Fall 1992).

finding support for the students."[45] Mukerrem Cakmak, one of James White's students from the University of Tennessee, came with him to Akron to serve as the chief engineer of the program.[46] Kyonsku Min-Cakmak (later Mukerrem's wife) also came to Akron in 1983. Although she finished her dissertation in Akron, she earned her Ph.D. in polymer engineering from the University of Tennessee in 1984 and became an assistant professor in 1987.[47] From its humble beginnings in 1983, the polymer engineering faculty has grown over the years and also remained remarkably stable. Kyu, Isayev, Cakmak, and Min are still members of a department which now includes twelve active professors and three emeriti.

The polymer engineering facility also grew through endowments. In 1992, the Dr. Benjamin Franklin Goodrich Endowed Professorship in Polymer Engineering was established. Chang Dae Han, director of the polymer science and engineering program at Polytechnic University in Brooklyn, was the first to hold the position. Goodrich chairman John D. Ong said that B.F. Goodrich himself had a deep commitment to technological excellence and scientific innovation, and the endowment would attempt to cultivate and honor his spirit in the university polymer program. "We are pleased," Ong said, "that a scientist of Dr. Han's stature will be the first to hold the professorship."[48]

While working to build the program, White also found time to advance his own research, which included attempts to process rubber to improve tire performance. He also studied twin screw extrusion, continuous mixing, and reactive extrusion. White received numerous awards for his work, including the 1981 Bingham Medal of the Society of Rheology, the 1984 Yuko Sho Award of the Society of Rheology of Japan, and the 1992 International Research Awards of the Society of Plastics Engineers. Nonetheless, he considered the highlights of his career to be his role as the founding president of the Polymer Processing Society in 1985 and his work as editor-in-chief of its main journal, *International Polymer Processing*. At a meeting of the American Chemical Society, White, and Lloyd Goettler, who was also from Akron, pushed for the establishment of the Polymer Processing Society. While there had been a Society of Plastics Engineers in existence since the 1940s, White and others believed its orientation was too American, and they wanted to form a new organization, more international in its scope. To cultivate the international ties, White's new Polymer Processing Society rotated its annual conference each year through three main geographic areas. One year it would be in the Americas, the next in Europe, and then the next in Asia, before rotating back to the Americas. The new organization was a success. To celebrate its tenth anniversary, it held its annual meeting in Akron. Over the course of three days, polymer researchers gave over 260 technical presentations on topics such as compounding, extrusion, injection molding, structure development, and blow molding.[49]

The recruitment of graduate students was also a key to the growth of the polymer engineering program. One distinguishing characteristic of the entire polymer program has been the number of international students. In 1990, one out of every three doctoral degrees were awarded to an international student. According to Frank Kelley, in the early 1990s, approximately 50 percent of the polymer science students were not born in the United States, and this increased to 90 percent of polymer engineering students. In 1992, the combined polymer programs had doubled the number of its graduate students, but at the same time had 24 percent fewer students born in the United States.[50] Because such a high percentage of international students were in the polymer engineering program, James White carefully considered what this meant not just to the university, but also to the country at large. History was, in a sense, repeating itself, he said, because in the past the textile and steel industries were built with the muscle of European immigrants. The difference now was that the influx of émigrés was coming from Asia.

White, always pleased to welcome these students into his program, cultivated and promoted its international focus. But there was a problem. It was not the high number of international students; it was the apparent disinterest of American students in attending graduate school in science or engineering. White said, "It's unhealthy if the country is not generating people who are strongly interested in technology, engineering and science because this is the backbone of the economy. . . . It's wonderful that, say, a South Korea has so many promising young people with this sort of enthusiasm, but it is disturbing that the same

Dean Frank Kelley (bottom row left) and James White (bottom row third from left) with a group of University of Akron polymer alumni from Korea at the IUPAC Symposium on Macromolecules in Seoul, Korea. *Polymer* (Fall 1996).

sort of enthusiasm doesn't exist among Americans."[51] The result has been a continuing decline in the number of students moving into graduate school in the sciences in the United States. While White was pleased with the trend of foreign students coming to the United States to study, he knew it was important to America's future to generate interest among those who would stay in the country. To encourage more Americans to enter polymer programs, the University of Akron began offering scholarships and fellowships open only to American students.

Akron's polymer engineering program has continued to have strong international participation. In 2004, the department had eighty-eight students from eighteen different countries. Since 1983, its graduates have gone on to make important contributions in industry and academia throughout the United States and the world. Some of these positions internationally include senior vice president and head of R&D at Kumho Tire in South Korea, professor at Tunghai University in Taiwan, and chief executive officer at Ram-Nath in India. In the United States its graduates are now president at Alcoa Flexible Packaging, a vice president at Teknor Apex, and head of the Polymer Science and

Polymer Engineering Branch of the U.S. Naval Air Warfare Center.[52] What unites these alumni is that they are using their polymer engineering education to further their own careers and also shape the future of their discipline.

In a remarkably short time, White and his colleagues had founded a polymer engineering program, established facilities, attracted funding, equipped the facilities with donations, and developed strong international student interest. At the same time he was looking for new ways to apply and use the polymer research generated by his program and support the polymer industrial community.

USING POLYMER ENGINEERING

There has always been a reciprocal relationship between the polymer industry in Ohio and the polymer engineering program at the University of Akron. After its first decade of operation, the College of Polymer Science and Polymer Engineering was placing 40 percent of its graduates with Ohio-based polymer companies.[53] One way this relationship grew was through the EPIC organization. White and Kelley advanced the University of Akron's participation in EPIC from the Olson Research Center, particularly through the work of the EPIC-Hanna Polymer Blending and Compounding Center, which became something like a small factory, transforming polymer research into commercial products.

One of the essential aspects of polymer engineering is blending and compounding. Before polymers can become finished products, they must pass through two important stages. Blending prepares polymers for their final uses, while compounding ensures that they are created with the right ingredients. When these two processes are correctly performed, the resulting polymer will exhibit the intended physical properties. In October 1989, the University of Akron announced the establishment of a new research center, which would become one of only a few of its kind in the world. The M.A. Hanna Company of Cleveland was the initiative's primary backer, and at the announcement ceremonies company officials presented a giant plastic check for $70,000 to President William Muse.[54] The Hanna Company later donated

In 2000, the Huntsman Packaging Corporation donated new equipment to the Institute of Polymer Engineering that James L. White said would "give us some unique research facilities." This included the ability to demonstrate multilayer film packaging. *Polymer* (Spring 2000).

another $120,000 to complete funding the center. Matching funds totaling $190,000 came from EPIC, and it also guaranteed another $180,000 in operating funds over three years.

The EPIC-M.A. Hanna Polymer Blending and Compounding Center was located in thirty-five hundred square feet of space at the Olson Research Center. Its mission was to enable basic research on mixing, reactive extrusion, and mixing machinery, and focused upon ways to determine if the ingredients were being used to match the intended physical properties of a polymer product. It also supported the polymer industry with technical services. According to Hanna president Douglas J. McGregor, his company was interested in this initiative because "The lack of talent in this industry is frightening." He hoped the students trained at the center would graduate and later help support polymer engineering needs in industry. James White emphasized, "We're in an age when blending and compounding have become increasingly important."[55] The Blending and Compounding Center became a place where researchers could objectively evaluate compound-

In 1993, Aubert Coran was the manager of the Edison Polymer Innovation Corporation (EPIC)-M.A. Hanna Polymer Blending and Compounding Center at the University of Akron. Here he examines a new $190,000 Buss America kneader machine used to mix polymers. *Polymer* (Winter 1993).

ing machinery donated by industry, including a twin screw extruder, a Banbury mixer, an internal mixer, and a two-roll mill. A Molding Technology Research and Development Center was also established in 1990 through the support of the Ohio Board of Regents to search for new and more efficient ways to process finished products through molding. These centers were essential due to the vital role of blending, compounding, and molding in polymer engineering.

In the early 1990s, the polymer engineering faculty continued to look for other innovative ways to use and apply their craft. In 1991, the university, along with corporate sponsors such as Alcoa, Instron, and 3M, worked to establish a Mechanical Testing Laboratory for polymer engineering. Alan Gent, who took on the project's administrative duties, believed this represented "a dramatic step forward," as it enabled polymer engineers to test the strength and stiffness of polymer products exposed to a range of temperatures.[56] In 1992, White announced another breakthrough development at the University of Akron. This was a software program called Akro-Co-Twin Screw, which was used to design

Professor Alan Gent testing physical properties of a dumbbell sample of rubber in an environmental chamber. College of Polymer Science and Polymer Engineering Archives. *Photo by Bob Wilkey.*

co-rotating twin screw extrusion systems essential for polymer processing and used in such key procedures as blending and compounding.[57]

In 1992, the *Cleveland Plain Dealer* profiled the numerous ways that polymer engineers at the University of Akron were "finding uses" for their research. For example, White Rubber, an area polymer company, developed rubber gloves used by utility workers. But they soon discovered a critical problem. At the time, they were making over one hundred thousand per year but found that, when the gloves were submersed in water, ten thousand volts of electricity would disintegrate portions of the rubber. Edward Kielar, White Rubber's technical services director, said, "Being a small company with a small lab, we didn't have the correct tools to diagnose the problem."[58] The company contacted EPIC, and the problem was solved in four days at one of the polymer engineering laboratories at the Olson Research Center.

The success of the polymer engineering program represented a bright spot for more than the University of Akron. During the 1980s, the city had faced numerous problems stemming from the loss of tire manufacturing and the virtual abandonment of downtown, but with the new decade came new opportunities and optimism. One report described the city as ready for a takeoff: "Downtown Akron is on the move—the sounds are everywhere. Jackhammers and arc welders, earth movers and wrecking balls, bulldozers and bellowing construction workers."[59] The first private downtown office building in the city in several decades was in the midst of construction. The new John S. Knight Convention Center was underway. The University of Akron also played a role in this development by taking ownership of the vacant Polsky's Department Store and converting it into new classroom spaces.

The downtown area took advantage of decaying rubber factories and its close proximity to the College of Polymer Science and Polymer Engineering. In 1994, Advanced Elastomer Systems (AES) looked at a crumbling former B.F. Goodrich factory and thought it would make a perfect structural and symbolic home for making new polymers.[60] The building, just south of downtown on Main Street, had been used by Goodrich to manufacture rubber bands, zippers, and the first space suits for NASA. AES, formerly located in St. Louis, became a leader in the polymer field, producing a thermoplastic elastomer known as

Santoprene.[61] William Ginter, AES's vice president for operations, said, "We've seen what is occurring in downtown Akron. . . . This building allows us to capture a unique climate." The company had been founded in Akron in 1982, so the move was a homecoming. Ginter said, "Consolidating our operations in Akron will help . . . [give] us closer proximity to the polymer science expertise of the area's business and academic community." Kelley described AES's return to Akron in one word, "Encouraging."[62] At its opening in February 1996, the new facility looked more like an art gallery unveiling as opposed to a former graffiti-covered, asbestos-laden, decaying symbol of Akron's fall from grace. Guests ate pastries and listened to the Roland Paolucci jazz band. The walls were lined with black-and-white still images from professional photographers chronicling the building's renaissance.[63]

Like AES, a number of other smaller polymer companies in Akron's industrial parks focused on polymer applications. They also relied upon the polymer engineering talent produced by the university. Several of the corporate residents of Akron's Ascot Industrial Park are thriving polymer companies. Tech Pro, which was profiled in *Wheels of Fortune* as an example of the rubber industry that remains in Akron, has been in business since 1983.[64] Though it is one of the world leaders in the manufacture of rubber testing equipment and software, it ironically got its start refurbishing old machines left over when older rubber factories closed down. A recent polymer engineering graduate from the University of Akron, Julie Orlando, serves an important role in technical services there. Just next door to Tech Pro, Herb and Ken Keuchel run Spunfab Adhesive Fabrics, founded in 1988 when this father-and-son team developed a specialty polymer adhesive fabric. Ken, the son, holds a master's degree in polymer engineering from the University of Akron, which he earned in 1994. Together they developed a product used for bonding textiles, leather, polyester, glass, and metals. Ken said, "The markets are very niche-oriented, and we make use of our polymer and engineering background."[65] Inter-Ion, just down the street from Spunfab, was established in 1994 by Adam Antonas (a doctoral student in the polymer program at the University of Akron) and Panos Panayiotu (who held a master's in mechanical engineering from the university).[66] The company creates coatings to protect metal parts. Other polymer companies at Ascot Industrial Park include HB Chemical, Graphic Arts Rubber and Win Plastic Extrusions. Main Street Gourmet, a bakery located in the middle of this complex, appears to offer a product as far removed from the polymer world as possible. But even its founders use polymer terms like "extruders" and "molding" to describe their high-tech bakery equipment. And it, too, has a close connection with the university as one of its founders, Harvey Nelson, serves as a member of the Advancement Council for the College of Polymer Science and Polymer Engineering.

As the exodus of the larger rubber companies continued out of Akron, the economic significance of these small companies increased. In 1998, B.F. Goodrich cut back its ties to the area. Although it kept its headquarters in nearby Richfield, Ohio, it moved a significant portion of its operations to Charlotte, North Carolina. It did, however, leave a chemicals and materials division in Brecksville, Ohio. GenCorp was also on the move. Founded in Akron in 1915 as General Tire, it moved its name and headquarters to California in 1998, including its defense, automotive sealant, pharmaceuticals, and aerospace work. It left its profitable polymer business, which makes graphic inks, latex products, coated fabrics, and vinyl wall coverings, in Fairlawn, Ohio. Frank Kelley was enthusiastic about GenCorp's decision to remain close to Akron, saying, "That decision said to me, this is where they belong if they're in polymers."[67]

Despite these advances, polymer engineering at the University of Akron still remained in the shadow of its scientific counterparts in one important respect—its facilities. A new polymer engineering building constructed on campus in the late 1990s strengthened the university's relationship with local industry and boosted future opportunities for the program.

THE POLYMER ENGINEERING ACADEMIC CENTER

By the mid-1990s, neither Frank Kelley nor James White was content with the polymer engineering home, and it became clear that the university needed to consider an expansion. The Olson Research Center's major deficiencies included spartan faculty offices, limited student workspace, and inadequate teaching facilities. Most of the students and faculty walked several blocks to the Goodyear Polymer Center for classes. Some said the polymer engineers felt like "poor cousins," shuttling between the new, modern science building and their renovated warehouse with its original 1920s construction. The polymer engineering program was also growing rapidly. In 1997, the program's size had increased by 40 percent over the previous four years.[68] The current facilities could not support this growth rate, and even worse, they threatened to inhibit recruitment. Kelley said that when prospective faculty and graduate students visited the university, they saw what the science program had and what the engineering program did not have. This was a "very, very crucial problem."[69]

White and Kelley decided the polymer engineers needed a modern facility they could call their own. Two important financial contributions supported their goal. In 1996, Sundar L. Aggarwal, director of the GenCorp research division, died, leaving $331,000 in his estate to the polymer program. One year later, the program received another significant financial contribution when the Rubber Division of the American Chemical Society presented the university with a symbolic check for $1 million. The chairman of the Rubber Division, Marshall Dean, said "The University of Akron has played an important role in the history of American rubber science and technology, and it plays a very important role today."[70] The check represented funds that became a ten-year lease for expanded headquarter facilities on the fourth floor in the new building.

The university officially broke ground for the new Polymer Engineering Academic Center in 1999. At the groundbreaking, President Luis M. Proenza said, "The completion of this building is the first step in a process of markedly enhancing the University's commitment to, and its standing in the field of polymer engineering."[71]

Two years later, on June 8, 2001, the Polymer Engineering Academic Center was ready to open with the unraveling of the giant blue-and-gold (university colors) polymer bow that had been wrapped around it. It was the first component of a $200 million campus makeover program called the "Landscape for Learning." Designed by the architectural firm Braun and Steidl, the new polymer engineering

Students enjoy the colorful main lobby atrium in the Polymer Engineering Academic Center. (From left to right are Wei Zhao, Todd Lewis, and Shanshan Wang.) *Photo by John Ashley.*

building's four-story atrium complemented the open spaces of the Goodyear Polymer Center; however, the new engineering building was smaller at thirty-five thousand square feet, and less expensive, at $6 million.[72] It was faced with glass and red brick. The Olson Research Center remained an important part of the polymer engineering facilities, thanks to its processing laboratories and heavy equipment. A second floor walkway connected the two buildings. The new structure also included a 134-seat auditorium, a forty-seat classroom, conference room, student computer laboratory, seminar room, and twenty-four workstations. There were twenty faculty and fourteen student offices that included modular furniture that accommodated twelve students per room.[73]

Frank Kelley believed its opening would mark a new era. Polymer engineers, he said, often "feel like they are ignored," but with the new building, this would no longer be the case; the new structure offered them "their chance to shine."[74] The opening was also a good moment to reflect upon how far the program had come in a short time. In 1983, the polymer engineering program had begun with just four students and one faculty member. In 2001, it had ten faculty and 110 students. The new building would be a place for advanced engineering research and training. It would also serve as a tool to recruit the best and brightest students and faculty.

The era of the planning and construction of the Polymer Engineering Academic Center was also a time for new leadership in the program. In December 1997, James White stepped down as department chair, and in December 2001 he also relinquished his duties as director of the Institute of Polymer Engineering. He remained at the university in an active research role, becoming the Dr. Harold A. Morton Professor of Polymer Engineering.[75] In 2000, Lloyd A. Goettler became its permanent chair, and one year later he was appointed director of the Institute of Polymer Engineering. In 2004, Sadhan C. Jana took over the chairmanship from Goettler. While he has only been in Akron for eight years, Jana said that the new building was a monumental achievement for the university: "It brought the polymer engineering program into prominence."[76]

The Polymer Engineering Academic Center is also the home of the educational outreach programs sponsored by the Akron Global Polymer

Sadhan Jana. *Department of Polymer Engineering.*

Academy (AGPA). Established in 2001 with a grant from the state of Ohio and a $1 million endowment from Goodyear, the academy is headed by the Goodyear Professor of Polymer Engineering. The first to hold this position was R. Byron Pipes, formerly president of the Rensselaer Polytechnic Institute, and director of Omnova Solutions in Fairlawn, Ohio.[77] Pipes, whose own research is in nanocomposites, arrived at the university in October 2001. The AGPA's primary mission is to teach the world about polymers at all levels of the educational spectrum. It uses distance learning techniques to support K-12 science instruction, graduate education, and workforce development. It maintains an interactive web site capable of conducting remote laboratory experiments.[78] Its course offerings include the Polymers Online Workshop for Chemistry Teachers; Rubber Lab Mill Safety Training; Ethics and the Professional Scientist; Experimental Mechanics of Composite Materials; and Ethics in Science for Science Teachers. Pipes said, "I love being a part of pioneering adventures. . . . It has been my life's work to focus on creating interactions between academe, industry, and the public."[79]

Key laboratories remain in the Olson Research Center. They include Kevin Cavicchi's block copolymerization laboratory, where he studies fuel cell membranes and researches new ways to develop tiny, inexpensive polymer computer chips.[80] Mark D. Soucek heads up a special coatings laboratory for a variety of products such as latex, polyesters, and powder coating, whose applications include aerospace, automotive, and floor coverings. His coatings are currently being tested for the new Patriot missile and for satellites. "Coatings are ubiquitous," Soucek says, "and we do it all here."[81] Mukerrem Cakmak's Wright Center of Innovation is also located here.

The National Science Foundation has also supported another key area of polymer engineering research, with a $333,134 grant to Avraam Isayev in July 2007. His work will focus on development of a continuous, ultrasonic process for the fast dispersion of nanofibers and carbon nanotubes in polymers. If successful, it will replace the current inefficient batch solution process to irradiate and remove solvents. The resulting lightweight nanocomposites could be used to enable higher speeds and larger payloads for jet fighters and commercial aircraft. "By selecting the University of Akron for this grant," wrote U.S. Representative Betty Sutton (D-OH), "the National Science Foundation recognizes the high quality of the work the university's researchers and academics conduct. I am extraordinarily proud of their hard work and their continued success."[82]

The diverse activities housed in the Polymer Engineering Academic Center and the Olson Research Center (including the Wright Center for Innovation, the AGPA, Rubber Division, faculty research, and student training) demonstrate that the polymer engineering program has

The Polymer Engineering Academic Center. The Sidney L. Olson Research Center can be seen at the far left behind the building. *Photo by Mark D. Bowles.*

finally achieved parity with the polymer science program. A showcase for recruitment, it promises that the university can fully realize the true potential of polymer engineering first imagined by its early, ardent proponents in the 1960s.

CONCLUSION—A THRESHOLD OF PREEMINENCE

Reflecting upon the growth and evolution of the polymer engineering program since 1983, Lloyd Goettler, the director of the Institute of Polymer Engineering, said, "One has to be impressed by the current stage we are in with our academic programs and our research."[83] At the beginning, the department focused on the structure of polymers and was funded by the rubber industry. This was the traditional classical interest of polymer engineering. Over time, the research mission evolved to become more focused on newer technologies such as nanoscience and nanocomposites, currently the province of the Wright Center of Innovation in Nanomaterials. The resulting applications are now different. Rather than creating the traditional consumer products of the past, polymer engineering's future lies in new arenas—aerospace, biotechnology, or smart materials that respond to the environment. Goettler predicts, "This will be the springboard for our program and our college. It will attain even greater eminence in the future." Akron's polymer engineering program, according to Goettler, stands at the "threshold of preeminence" in relationship to other polymer programs in the nation. Nearly a quarter-century after its founding by James White, polymer engineering has come of age. It has shed its status as a poor cousin to the polymer science program on campus and is earning recognition as one of the leading polymer engineering academic centers of the world.

7

THE POLYMER STATE

1999–2007

IN APRIL 2001, *NEWSWEEK* PUBLISHED AN ARTICLE celebrating the emergence of a "new brand of tech city." Journalist Adam Rogers and a team of reporters identified a list of cities building new economies with unique innovation strategies.[1] The first city mentioned was Akron, Ohio. Rogers retraced the story of how Akron "smelled like molten rubber" through most of the twentieth century, only to have most of its major rubber companies pull up their stakes in the 1970s and 1980s. This left thirty-five thousand manufacturing jobs lost in its wake, but something important remained—a polymer "research infrastructure." *Newsweek's* journalists suggested that Akron's dedication to the field of polymers would enable it to play a vital role in future commercial ventures as well as stimulate a possible economic renaissance. The University of Akron's polymer program was the intellectual center of this research infrastructure. It not only graduated the leaders in the polymer field, but it also supported the growth of four hundred polymer-based companies in northeast Ohio that employed thirty thousand people. Akron Mayor Donald Plusquellic, who was very enthusiastic about the *Newsweek* recognition, said, "There's nothing much more high tech than polymers and changing the molecular structure of products so as to improve [them]. . . . That's going on right here in a major way in Akron."[2]

Despite this recognition and optimism, significant problems were brewing under this high-tech polymer exterior. The late 1990s was a time of change in the polymer industry, including a redirection of state support for polymers and turmoil at the head of the university itself. Akron's recent inabilities to retain the headquarters of many of its large rubber companies required forging a new direction, one increasingly shaped and defined by smaller engineering companies. If Akron was to continue to prosper and capitalize upon its polymer and rubber heritage, it needed to cultivate an environment where these smaller polymer companies could grow and prosper. There would be significant consequences, including EPIC's loss of state funding, and its eventual demise.

But there were deeper problems on the horizon than the loss of EPIC as many questioned the state's political leaders' commitment to and understanding of the important role polymers played in the economy. For example, ten times more funding was available for coal research (which represented less than 1 percent of the state's revenue) than polymers (which represented 6 percent of the state's revenue).[3] What was required were efforts to address this imbalance and prevent Ohio from losing its polymer leadership position in the national economy. Out of this troubling period emerged a new voice at the University of Akron—Luis Proenza, the first university president since Hezzleton Simmons to have been a scientist before becoming an administrator. His skills, energy, and enthusiasm would help Dean Frank Kelley steer the polymer program through a difficult period and lead the struggle to advance the state of polymers in Ohio. These efforts are ongoing today.

PROENZA'S CHALLENGE TO OHIO

As the economy rises and falls, so, too, are there peaks and valleys in the funding sources for large institutions. In the late 1950s, when the University of Akron was still a municipal school, it suffered from declining tax revenue from the state. Maurice Morton encountered another such period in the early 1970s when he rallied against funding cutbacks by the National Science Foundation, which had forced the university to terminate several chemistry and research programs.[4] Morton, testifying before the congressional House Subcommittee on Science, Research and Development, said, "Drastic reduction cannot but have a catastrophic effect on education in science." He told the legislators that a decrease in funding research, would "be to the detriment of our country's progress."[5]

While the university increased funding in the 1980s, scarce resources became a central reality of President Peggy Gordon Elliot's tenure a decade later. Elliot became president of the University of Akron in August 1992, succeeding William Muse.[6] She appeared well qualified to guide the University of Akron, having previously been chancellor of Indiana University Northwest for nine years and having written a book on urban campuses.[7] The main problem she faced in Akron was a decline in financial support from the state, decreasing enrollments (nearly 20 percent during her tenure), and discord among the faculty. She responded by cutting $20 million from the budget and reducing faculty.

The problems only intensified a few years into Elliot's tenure when concerns about the university's governance, and, more specifically, its president, arose suddenly in January 1996. Sixty-five members of the faculty senate questioned Elliot's hiring practices and her alleged nomination of faculty members for positions at other universities without their knowledge. The Faculty Senate asked the Board of Trustees to examine the situation, and they in turn passed the matter to Betty Montgomery, Ohio's attorney general.[8] Montgomery appointed Cleveland lawyer Thomas Wiencek as a special counsel to investigate the situation.[9] This was a difficult period that affected all the university's academic programs, including the polymer program. In February 1996, the members of the Advancement Council of the College of Polymer

Science and Polymer Engineering wrote an open letter to the Board of Trustees requesting that closure be brought to these problems as soon as possible. The council worried that ongoing public concerns would damage the reputation of the polymer programs and would potentially lead to diminished funding among industrial supporters. They wrote, "Continued public acrimony within the University community will only adversely affect the global future of the college."[10] But the disagreements and fighting among the Board of Trustees could not be easily resolved. On April 2, 1996, Peggy Elliot resigned. Marion A. Ruebel was appointed interim president, and a search to find a permanent replacement began.

This search ended with the hiring of Luis Proenza, a person who blended the talents of both scientist and administrator. Proenza had led a geographically diverse academic life prior to coming to Akron. He was born in Mexico City in 1945, and at age eleven his family sent him to the United States to attend the Riverside Military Academy in Georgia. He attained the highest student rank possible there—Battalion Commander—thus beginning what he called his "formative stages of leadership skills."[11] He attended Emory University, where he became interested in the science of the mind, earning his bachelor's degree in psychology and philosophy in 1965. A year later, he enrolled at the Ohio State University, where he completed his master's degree in psychology. There he studied abnormal behavior and the inability of schizophrenics to control sensory input to their brains. He continued his research at the University of Minnesota, where he earned his Ph.D. in neurobiology in 1971.

After completing his studies, he became a faculty member at the University of Georgia, where the National Eye Institute supported his work. He excelled in his research, becoming full professor of zoology after sixteen years. He also became actively involved in administration and was the study director for the National Academy of Science's Committee on Vision, in Washington. He discovered, through his experience in Washington, that he greatly enjoyed program building and policy matters. In 1987, he found an outlet for this interest at the University of Alaska, where he became the vice president for academic affairs and research, and later vice chancellor for research and dean of the gradu-

University of Akron President Luis M. Proenza. *University of Akron, Office of the President.*

ate school. He remained in Alaska until 1994, when he moved to Purdue University to become vice president for research, dean of the graduate school, and professor of biological sciences.[12]

Four years later, a presidential search committee contacted him from a university he had never heard of before. Although he had spent a year working on a graduate degree at Ohio State, his knowledge of the University of Akron was strikingly limited prior to his arrival. When asked about the school, he said, "I didn't know it existed." But once the search committee had contacted him, his interest was piqued. He said that when they told him that the polymer program ranked in the top two nationally, "You pay attention." The committee liked Proenza's enthusiasm and passion and offered him the presidency. Looking back on his arrival at Akron, he said, "No university president . . . has a right to expect or let alone actually find as much excellence as I have found here."[13] Named the University of Akron's fifteenth president on October 14, 1998, Proenza officially began his tenure as president at his investiture on January 1, 1999.[14]

Proenza confronted some immediate challenges. Enrollment was declining for nearly all universities in the state, but by the time Proenza took the helm, Akron was the only institution still unable to recover. Proenza believed the Akron campus "still looked a little bit too much like just a couple of city blocks of not particularly distinguished buildings."[15] Except for the Goodyear Polymer Center, it had been several decades since the university had constructed a new building. Proenza said students were "voting with their feet" and going to schools with modern amenities like new student unions, residence halls, and recreation centers. It was time to start bringing them back to Akron.

In response, he initiated a $200 million "Landscape for Learning" plan whose goal was to double the open green spaces on campus and create an inviting place for new students to visit, stay, and become educated. His plan also included physical facilities such as a student center, an honors dormitory, a recreation facility, and a significant expansion of the performing arts building. In remaking the physical campus, Proenza and the university architects borrowed the primary aesthetic from the Goodyear Polymer Center's sloping roof structure, glass front, and silver steel accents. Even the new parking decks emulated this look with their corner, winged glass structures. Proenza also took on the responsibility of chief spokesman for the school, heading an over $1 million television advertising campaign (the largest in the institution's history) to reintroduce the university to the local community. The success of Proenza's master plan was to be measured both by the number of new students and also the amount of new funding he could attract. He said, "If we don't improve our physical plant, if we don't manage our enrollment, we won't have dollars coming through the door," dollars that came primarily from tuition and state funding.[16] The first "Landscape for Learning" building constructed was the Polymer Engineering Academic Center, which was dedicated in 2001.

Funding also remained a top priority for the polymer program. Shortly after his arrival, the *Cleveland Plain Dealer* ran a column by Proenza assessing the state of Ohio in terms of its scientific and research potential. He said that by size measures alone, Ohio was positioned well, with the seventh largest state economy in the nation. But he criticized state leaders, suggesting that Ohio had become complacent while Georgia and North Carolina took bold steps, cultivating emerging industries and strengthening their institutions of higher education. By investing in their research infrastructure, these states began to attract massive amounts of federal and industrial funding. Proenza said, "By comparison, Ohio and other states in the Midwest have lagged in innovation and lost market share in R&D."[17] Ohio did not have a public university ranked in the top 25 percent nationally, and only three were in the top 50 percent.

Proenza was adamant that the state had to do much more to return to its former place of prominence. The lone bright spot was the strength of the polymer academic programs in Ohio. But here, too, Proenza was concerned. He said, "Even in the $20 billion polymer sector, Ohio's investment to sustain competitiveness seems paltry."[18] From Proenza's perspective, the meager state support of polymers represents a significant threat to the intellectual and economic vitality of the region. In 2002, he told a Cleveland City Club gathering, "Rather than benign neglect, this paltry support borders on criminal negligence."[19]

His example of this was the Edison Polymer Innovation Corporation, which represented Ohio's major source of funding for polymer research activities. Compared to its large polymer industry, the state only provided roughly $1.2 million per year of funding to the EPIC program. Polymers represented 6 percent of Ohio's gross state product, yet the state invested less than a thousandth of a percent of the polymer industry's total output. By whatever statistic used, polymers were greatly underfunded. In 1999, Proenza called for a "commitment so bold as to signal strategic intent from the state."[20] Shortly thereafter, in sharp contrast to Proenza's challenging rhetoric, the Ohio Department of Development ended its support of EPIC, its most visible and at one time most optimistic polymer initiative.

THE ENDING OF EPIC

In 1993, even while the promises of a Polymer Valley seemed assured, Frank Kelley warned of "clouds on the horizon." The University of Akron had the largest university-based polymer research facility in the country. Ohio ranked number one in plastics, an industry that employed more than one hundred thousand Ohioans. Nonetheless, the state offered little support for its polymer programs, in contrast with other states—Illinois, Louisiana, and Michigan for example. According to Kelley, "Part of the problem is that as a state we have not yet acknowledged our position as leader in all aspects of polymer technologies. . . . Our state government is just waking up to this fact."[21]

This same message was reinforced one year later at MacroAkron '94. This event brought seventeen hundred polymer researchers from forty-five countries to the University of Akron to give technical presentations, examine the challenges in the polymer industry, and look toward the future. This was the thirty-fifth biennial symposium sponsored by the International Union of Pure and Applied Chemistry, but it had not been held in the United States for twelve years. Kelley called the symposium a "chance of a lifetime."[22] Joseph Kennedy, an Akron polymer science professor, chaired the symposium and had been working on it since receiving approval for the event in 1988 at a previous meeting of the group in Kyoto. Kennedy took great pride in hosting the conference, which represented the largest research gathering in the city and potentially the most significant polymer assemblage ever in the nation. "This is a big deal," he said proudly.[23] Although state politicians, including Governor George Voinovich, Congressman Tom Sawyer, and Mayor Don Plusquellic, all spoke in glowing terms about the financial impact of polymers, keynote speaker James Economy, a polymer scientist at the University of Illinois, cast some doubt on the issue. Basic research, he said, was threatened by industrial budgetary cutbacks. Organizations were hesitant to adopt any new technology that could not be easily transformed into a profit. He said that academics needed to understand the marketplace better, and industry and universities needed to find new ways to work together.[24] Government might play a key role in making this happen.

A computer rendering of the Goodyear's Spirit of Akron blimp welcomes more than fifteen hundred scientists and engineers gathering for MacroAkron in 1994. *Polymer* (1994).

In an attempt to improve this situation, plastics executives from various organizations throughout the state traveled to Columbus in 1994 to meet with government officials to discuss possibilities for increasing funding and commitment to polymer growth. The meeting was encouraged by the plastics executives from companies such as Rubbermaid and sponsored by Governor George Voinovich and the Ohio Department of Development. There were some important indicators that the state was heading in the wrong direction. Ohio had a $15 billion plastics and rubber industry that led the nation. However, since 1987 its production in these areas had fallen when compared to the rest of the United States. Capital investment in the polymer industry declined against the national average. The goal of the plastics executives' meeting was to raise awareness of the problem and encourage state legislators to promote more aggressively an industry so important to Ohio's future.[25] The meeting perhaps increased awareness, but it did not bring immediate results.

With funding from government sources reduced, the College of Polymer Science and Polymer Engineering looked toward increasing its support from industry. One major gift came from the GenCorp Foundation in 1996. This $1 million gift, the largest in GenCorp's history, offered $200,000 initially, followed by six annual payments of $133,134. The funds were intended for an equipment endowment. The ability to purchase more advanced equipment would in turn allow the university to be eligible for more research grants. Russell Livigni, the vice president of corporate technology for GenCorp, said the endowment was created because the polymer college was "an important resource" and vital to the company's future.[26]

Despite the continuing close ties between the polymer college and the polymer industry, the looming crisis of uncertainty surrounding state support threatened the future of EPIC. In the 1990s, EPIC was enjoying some successes through such advances as biomimicry, the attempt to manufacture synthetic materials based upon polymers that existed in nature. One unusual project was a study involving golden orb spiders from Central America whose webs were stronger than nylon. The spiders lived in "condos" on the seventh floor of the Goodyear Polymer Center. Ronald Eby, the Robert C. Musson Professor of Polymer Science, headed the research. Understanding how the spiders made their natural polymer webs at room temperature promised advances in improved surgical sutures, bulletproof vests, and other polymer products. In a similar natural study, Case Western Reserve University examined another unusual natural polymer—reindeer antlers—because they were incredibly strong, lightweight, and fast growing. The research promised results from the profound (stronger medical implants) to the mundane (bowling balls that never chipped).

In 1992, EPIC's academic connections received high praise from Thomas Purcell, Ferro's vice president of research and development, who said, "Schools like Akron and Case Western Reserve University are important. Industry doesn't do much basic research anymore."[27] Ferro paid $40,000 per year in dues to be a part of EPIC and share in its research findings. EPIC was also expanding its academic affiliations. In

In 1992, the University of Akron polymer researchers studied the silk produced by the golden orb web spider in the hopes of improving polymer materials, like high-performance fibers, surgical sutures, and bulletproof vests. Dr. Ronald K. Eby and graduate student Dave Mahoney studied the spiders' dragline silk, which is three times stronger than silkworm silk or nylon. *Polymer* (Spring 1992).

1998, the University of Akron and Case Western Reserve University agreed to extend their partnership in EPIC to include Ohio State University. William E. Kirwan, the president of Ohio State, came to Akron to deliver an address reinforcing his commitment to the new relationship. William O. Overton, EPIC's president, recognized that Polymer Valley now stretched much farther south and said, "We're a lot bigger than a valley now. Maybe we'll have to call ourselves the Polymer State or the Polymer Continent."[28]

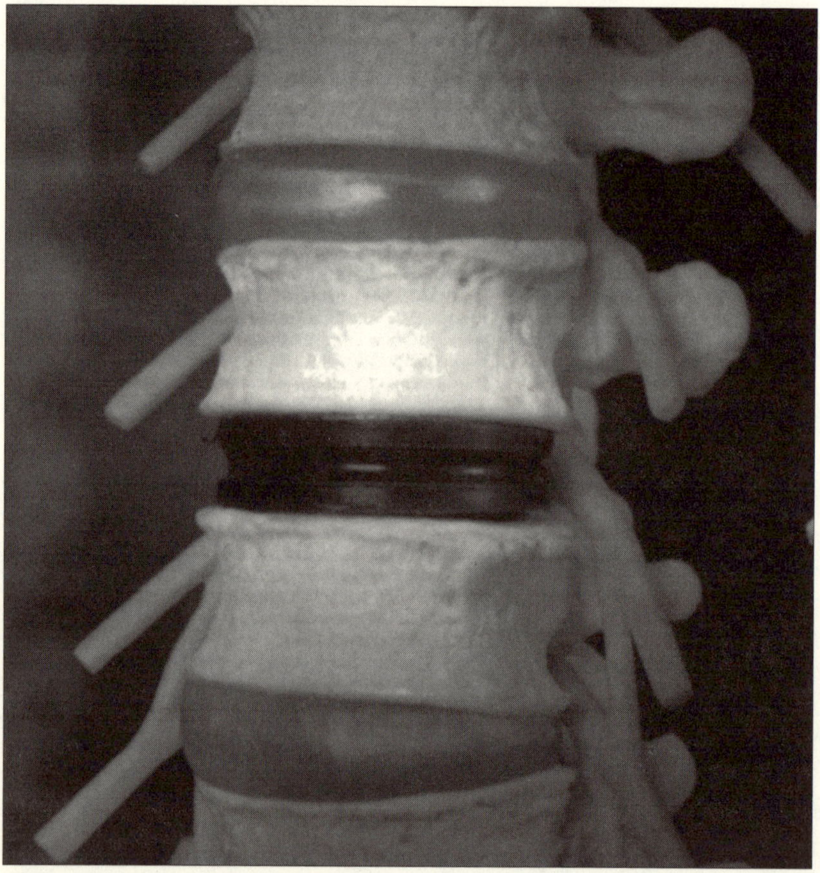

This model human spinal column is shown with an artificial disk developed by EPIC-member, AcroMed Corporation. The EPIC Mini-Pilot Plant at the University of Akron synthesized its base polymer. *Polymer* (Fall 1996).

Another new EPIC initiative in the late 1990s was a consortium of industry and academic partners working to create new polyvinyl chloride (PVC) materials. These had previously been used for products such as house siding, windows, and automobile parts, but there had been no major advances in the use of these materials for two decades, and PVC's usefulness was threatened by other, newer polymers. The consortium's goal was to develop new PVC materials with improved qualities for applications that could potentially open up new markets.[29] The project's financing came from the companies involved in the consortium. For $125,000, they could receive patent rights. If they wanted access to the research, the cost was $15,000 to $25,000. The research took place at the universities. Once new products were licensed by the corporate sponsors, the universities received a share of the revenues. The main problem with PVC was competition abroad and the low profit margins at home. EPIC's leaders believed the consortium's coordinated research would enable the discovery of PVCs with increased durability, strength, and temperature resistance.[30]

Despite these advances, the hopes and dreams of an EPIC-shaped Polymer Valley soon faded away. There were inherent tensions within EPIC that were becoming increasingly difficult to overlook, primarily the disconnect between the long-term goals of basic research at the universities and the short-term commercial needs of the member companies. Richard Celeste, the Democratic governor who helped establish EPIC, had been more tolerant of the time required for the application of basic research. In contrast, his Republican successor George Voinovich (governor from 1991 to 1998) and the Ohio Department of Development, which oversaw the EPIC program, wanted to see more immediate results. They wanted to define EPIC's success objectively, a nearly impossible measurement to make. A majority of the research was funded by large corporations, which then incorporated the results into their own internal research and development. It often took a significant period of time for a viable commercial product to emerge from this R&D process; according to Kelley, "That was one of the reasons why it was so difficult to show on the key performance indicators to the state how many new businesses were started."[31] EPIC represented a grand vision to create a Polymer Valley supported by large companies. While the dream remained, the state leaders believed they could use their funds more efficiently.

Jerome Lando, a professor of macromolecular science at Case Western Reserve University, said the Voinovich administration wanted an immediate accounting for EPIC's achievements. But Lando believed its contributions were impossible to measure directly because, he said, "Companies would take our research back in house, and wouldn't tell us what they were doing with the stuff." When the *Akron Beacon Journal* tried to report on how the local polymer industry used EPIC's research, the corporate sponsors declined to talk, pleading that it was proprietary information. Samir Gibara, the Goodyear chairman, said this was the major difference between professors and businessmen. "Professors like to publish the results of their research. We like to keep things quiet until we are ready to announce them." And it often took a long time in development before a company was ready to announce a new polymer product.[32]

EPIC's demise started in 1999 when the Ohio Department of Development, then under the control of Republican Governor Robert Taft, considered suspending its funding. In 1999, the total state funding for the program was $1.3 million, nearly one-half of EPIC's $3 million budget. According to Pat Valente, the deputy for the Department of Development's technology division, "We were dissatisfied. Polymers are probably [among] the top three or four industries in Ohio . . . but from our standpoint, EPIC just wasn't going in the right direction." A new direction, the state hoped, would create an environment in which start-up polymer companies would thrive. EPIC, with its focus on larger established organizations, was unable to do that. Frank Kelley said, "There was this dichotomy . . . a problem of trying to match the goals of start-up businesses with the funds from large corporate contributions."[33] In October 2000, Taft made the official decision to stop EPIC's funding.

Members of EPIC were shocked by the announcement. Tom Purcell, EPIC's chairman, said it was unexpected. He knew, however, that the state wanted to see more job creation, and EPIC was not achieving that goal. EPIC President William Overton said, "Our world has been turned on its ear."[34] The end of EPIC was a disappointment from the perspective of the University of Akron, but EPIC funding represented only between 10 and 15 percent of the College of Polymer Science and Polymer Engineering's external support. Despite the loss, university leaders could point to lasting accomplishments made by EPIC. Most important was a stronger relationship with Case Western Reserve University. EPIC helped to build an academic bridge between two renowned polymer programs, and the end of EPIC did not diminish this relationship.

In an attempt to replace EPIC, in 2001 the state created a new nonprofit organization, backed with $2 million in funding, called the Ohio

Polymer Enterprise Development, Inc. (OPED). Though some referred to it as the "little son of EPIC," it was actually quite different. Its goal was to help create at least six new polymer businesses a year, in the hopes they would quickly evolve into companies employing five hundred people with $50 million per year in total sales. Charles Wilkes, the group's interim president, led the effort to evaluate over eighty technology patents for commercial applications. Wilkes said, in July 2001, "We're looking for highly promising businesses, and we already have a large portfolio of possible ideas."[35] These included polymer fibers to bandage skin cuts and abrasions, polymers that would improve laptop screens, and polymer membranes physicians could inject into patients to release timely doses of insulin. Patents already existed and OPED's role would be to support the best ideas and provide assistance through the commercialization process. The University of Akron's polymer program remained closely involved with OPED, just as it had with EPIC. Frank Kelley sat on its board and also wrote the proposal that secured the $2 million grant from the Ohio Technology Action Fund. Another key component of Wilkes's job was to run a polymer pilot plant housed at Omnova Solutions. The equipment included five chemical reactors that enabled scientists to produce specialty polymers, which customers could then evaluate for use in new products.

Wilkes said the effective commercialization and creation of useful products from polymers required the freedom of time and the luxury of money. He said, "Any professor can make a gram of something in a lab, but to make it in tons and tons and tons at a price that people can afford to pay, that takes a lot of engineering work."[36] The costs required for this development increased exponentially. If a research project cost $10, it might require $100 to develop the idea, and another $1,000 to begin actually manufacturing it. For this reason, the State of Ohio believed it needed to step in to support the development of new polymer ideas and start-up companies. But Wilkes warned that, historically, nearly every major invention has required at least a decade between its original inspiration and commercial development. His warnings and request for patience fell on deaf ears.

In October 2001, Ron Clark took over from Wilkes as OPED president. He had earned a Ph.D. in polymer science at the University of

Akron, worked in research for Goodyear and General Tire, and later became vice president for marketing at Akron Rubber Development Laboratory. He had also established several of his own small companies. Clark said, "I've always enjoyed starting businesses and putting teams together and the capital behind them to make them successful."[37] One of the most successful OPED-supported businesses during Clark's tenure was Akron Polymer Systems, which was established in 2002. Spun off from research performed by University of Akron professors Frank W. Harris and Stephen Z.D. Cheng, they developed a polymer film coating that, when used on the surface of liquid crystal displays, dramatically increased the viewing angles. The company now works to develop and commercialize high performance polymers for aerospace, optical, and photonic applications. Harris said of Cheng, "Working together, we have greatly increased the significance of our accomplishments."[38] Today the company employs six full-time Ph.D. polymer science researchers.

Stephen Z.D. Cheng. *Polymer* (Summer 1994).

Other successes soon followed. OPED helped GriggSmith Industries attempt to commercialize wound-care technology. PolyDisplay was a Norwegian company lured to Akron by OPED to develop liquid crystal plastic laminates for car windows and billboards. Steve Johnson, a majority owner of Advanced Coatings International (which made polymer coatings for circuit boards), moved his company from Israel to "Polymer Valley" and sought new sources of capital with the assistance of OPED.[39] By November 2002, there were other encouraging results, and some observers calculated that OPED was responsible for thirty polymer start-up companies. Clark expressed caution, saying, "This is not an

Frank Harris. *Polymer* (Fall 1995).

overnight thing. But we are at a milestone now. We've created a strong foundation. Things will start happening very soon."[40]

Things did start happening—but not in a positive direction. Though OPED supporters always emphasized that it was different from EPIC, the two organizations shared one unfortunate commonality: both met their untimely demise due to a sudden termination of state funding. After two years, in April 2003, the Ohio Department of Development made the decision to terminate support for OPED.[41] The state was in the midst of a budget crisis and, as a result, planned to allocate no more resources to OPED after December 2003, when its existing funds ran out. In total, the state had granted it $2 million. The state decided that OPED was not attaining its goals, even though two years was a very short evaluation period and it already had one very successful company, Akron Polymer Systems, to its credit. Frank Kelley said, "It was unfortunate to have terminated it two years from the start and I believe very heartbreaking for Ron Clark, who was totally passionate about it."[42] Local opinion also weighed in against this decision by the state. An editorial in the *Akron Beacon Journal* stated that in this "detour to polymer valley" the state was taking a "wrong turn."[43]

In May 2003, ten OPED management consultants helping polymer companies in Ohio were terminated. One month later, OPED officially closed its doors, and its two full-time employees, a president and a secretary, were released. It also shut down the pilot plant at Omnova. For the rest of 2003, OPED remained an organization with no employees, managed only in name by the university. Journalist John Russell tabulated an OPED scorecard at the end of its life and said that, overall, the results were mixed. OPED had established, by his count, only eight new companies in northeast Ohio and twenty-five new jobs. But it made important contributions to advancing the general discussion on ways to commercialize new polymer developments.[44] And it had very little opportunity to achieve lasting success.

But, despite the turmoil, EPIC is not entirely dead. It survives today as a viable nonprofit organization. All its employees are now retired except one, Charles Wilkes. He remains on a contract basis because he still has a pool of funds totaling several million dollars obtained from industrial sponsors to support the PVC Technology Consortium. Currently, all of the research projects have concluded, and the efforts of Wilkes and the EPIC Board of Directors are focused upon securing patent rights and licensing the technology.[45] The hard lesson learned from EPIC and OPED is that funding is often ephemeral. Despite the inherent nature and time requirements needed to develop polymer products (or any invention), long-term funding continuity will always remain an Achilles heel for projects involving basic research. However, in the midst of the disappointment over the loss of these programs' support, there was some hope on the horizon with a new state initiative designed to invest in high technology initiatives. Could the polymer community capitalize upon these opportunities?

THE WRIGHT BEGINNING

University of Akron President Luis Proenza was not content to let the opportunities offered by EPIC and OPED slip away. Instead, Proenza said these setbacks resulted in a renewed attempt to "begin a really strong campaign for us to link better to the industry, and to put in place a new industry association."[46] The university was "not getting money anywhere from the State," and industry funding was not as strong as it should have been. Proenza concluded that "It was immediately clear that one of the things that the industry needed to do was advocate for itself." He began talks with Samir Gibara, chairman of Goodyear, and they invited more than ten CEOs of major Ohio companies to meet with them. This marked the beginning of the Ohio Polymer Strategy Council.

The council, formed in April 2001, represented a dynamic team of government, academic, and industry leaders with a shared mandate to strengthen Ohio's polymer community and also encourage greater state support. The council also served as an important advisor to PolymerOhio, a trade organization established to assist the polymer industry with promoting itself, directing its activities, and coordinating the diverse groups involved in polymer research and development. By some estimates, there were twenty-eight hundred polymer companies, employing 140,000 people in the state. These companies generated \$49 billion in revenues as well as \$5.6 billion in workers' salaries. PolymerOhio became an important organization whose goal was to become an "industry connecting and networking group that is focused on enhancing the global competitiveness and growth of Ohio's plastics, rubber and advanced materials industry."[47]

The council specifically assisted PolymerOhio with defining strategic initiatives; advising the governor's office about polymer developments and opportunities; stimulating federal and state investment in polymer research in Ohio; and striving to improve technology transfer between the academic and industrial communities. Representatives with ties to the University of Akron played important roles in this council and with PolymerOhio. Sharell Mikesell, who earned his Ph.D. in polymer science from the university in 1971, was the first executive director of the Ohio Polymer Strategy Council. Proenza also serves as a current member, with other prominent members of northeast Ohio's academic and industrial community.

Another Proenza initiative to attract more state support was the promotion of George Newkome as not only dean of the graduate school but also vice president of research, a position Proenza elevated to a much higher status within the university leadership structure.

Newkome emerged as a compelling candidate for this job. At the University of South Florida in Tampa, he worked for fourteen years toward increasing research funding from $20 million to $171 million per year. After Proenza hired him, Newkome also became a professor of polymer science at Akron, where his own research led him to investigate nanotechnology and the attempt to create molecular batteries and motors. One goal of his new responsibilities was to coordinate with local industry and assist in spinning off new companies from the university. Newkome said, "A lot of people are trying to be like Silicon Valley. And that's difficult. You can't take something that's been around for fifty years and say, 'I'd like to be like them.' That doesn't work. But we can be a focal point for polymers. We just need our own game plan."[48] Newkome is currently a member of the board of trustees of PolymerOhio.

Efforts to stimulate polymer funding from the state have succeeded in some areas, most notably in the Wright Center for Innovation. In the summer of 2002, Governor Robert Taft established a new initiative, the Third Frontier Program, an economic development program that focused on creating start-up technology potential. One of its central components was, initially, a $1.6 billion program, the Wright Centers of Innovation. These centers were established with a series of grants that supported "large-scale world-class research and technology development platforms designed to accelerate the pace of Ohio commercialization."[49] Participating in these Wright Centers were Ohio universities, nonprofit research organizations, and Ohio companies. The high technology areas that the state hoped to fund included advanced materials (such as polymers), bioscience, power and propulsion, information technology and instruments, controls and electronics.

The University of Akron began its efforts to win a Wright Center for its polymer program in June 2003, the same month OPED closed its doors. However, the initial attempt met with failure. The Ohio Department of Development contacted university officials and rejected the university's bid for $10 million in state funds to build a Wright Center of Innovation to support the polymer industry. It would have been called the Wright Center for Advanced Materials and Process Innovations. The state decided instead to allocate Wright Center funds to a fuel-cell commercialization project at Case Western Reserve University. But the state's deputy director for the Ohio Department of Development's technology division, Pat Valente, said, the university's proposal was "very good. There were a lot of proposals. The state had to say no to many of them this time. We're strongly encouraging them to reapply."[50]

Five months later, the university made another submission, this time for a proposed Wright Center for Polymer Photonics. Again, Akron lost out to other projects that appeared to offer better opportunities for creating new jobs. Polymer photonics was an important type of material that was capable of interacting with light in such a way that made it an excellent surface material for computer and television flat panel and liquid crystal displays. The evaluation committee criticized the Akron proposal because it was not more closely aligned with industry, explaining that "The proposed center appears to be more a collection of projects that would be located in a collegial academic environment, not a well-integrated organization that would contribute to job creation."[51] This time, the funding went to the University of Dayton, for an advanced data management center, and Ohio State, for a propulsion and power center. Some polymer experts believed Akron's problem was that the evaluation committee did not fully understand the impact or the importance of polymers. It was in some respects easier for nonscientists to appreciate such concepts as data management and propulsion, but understanding polymers required more specialized knowledge. University officials did not give up; they simply resolved to try to communicate their goals better once they were eligible to submit a new proposal.

Further troubling news came just two days after the second rejection. In the November 2003 elections, Ohio voters failed to pass Issue 1 by a margin of 51 to 49 percent.[52] This issue would have generated an additional $500 million for Governor Taft's Third Frontier program that funded the Wright Centers of Innovation. As a result, the program's overall funding had to be reduced from $1.6 to $1.1 billion. There would still be competition for new Wright Centers of Innovation, but there would be less money specifically earmarked for universities to develop new ways to commercialize high-tech products.

In August 2004, the university tried once again, with another detailed proposal, to win funding as a Wright Center for Innovation. This time there was a total pool of $65 million available; twenty applicants from around the state competed for funds. The university thought it had learned from its two previous failed attempts and structured its proposal to concentrate on job creation and industry partnership. It also broadened its focus to include nanocomposites, which incorporate nanoscale particles, fibers, or nano layers of thin films. The polymer is the matrix that holds the nano components together. They have tremendous potential in a variety of new applications from chemical circuits to biomedical instruments. Another new strategy was a partnership approach, where the University of Akron teamed up with researchers at five other public universities in Ohio. Frank Harris, a professor of polymer science at Akron, said, "We're optimistic we have a good shot."[53]

After three tries, in May 2005, the University of Akron, along with its academic partners, was awarded a share of a Wright Center of Innovation grant.[54] Total funding was to be $22.5 million to form the Center for Multifunctional Polymer Nanomaterials and Devices. Roughly $6 million of this was awarded to Akron, with the rest divided among the other academic partners. The Ohio State University (which led the proposal effort) received $7.5 million to work on biotechnology applications. The University of Dayton received $5 million to research ways to create stronger polymer composites with fewer imperfections. Smaller amounts were awarded to the University of Toledo ($1 million), Kent State University ($375,000), and Wright State University ($250,000).

Sharell Mikesell, executive director of the Ohio Polymer Strategy Council, believed the award would result in the creation of forty-five hundred high-paying jobs, twenty-five hundred lower-paying support jobs, and the retention of an existing twelve thousand jobs in the region.[55] Additionally, there were other cash investments and a match of $52 million for the new center by sixty polymer companies in Ohio. These included Goodyear, GE Transportation, Sherwin-Williams, Honda, Timken, Owens Corning, Ashland, Keithley Instruments, PolyOne, Goodrich, and Lockheed Martin. Mikesell said, "The announcement today is an exciting culmination of many months of effort by Ohio's polymer leaders, the Ohio Polymer Strategy Council, key university leaders and industry collaborators." The Wright Center, he believed, would "significantly strengthen Ohio's position as the world's

Sharell Mikesell. *Polymer* (Fall 1992).

premier location for the polymer and advanced material industries."[56] Mikesell and James Lee of Ohio State University would serve as codirectors. Proenza was equally enthusiastic. He said, the center would "provide the University of Akron a major opportunity to strengthen our world-renowned polymer program. . . . Our new collaborative efforts . . . will contribute substantially to Ohio's leadership in the global polymer community."[57]

On June 6, 2005, Mukerrem Cakmak became the site director for the new Wright Center for Innovation facilities at Akron.[58] The center was to develop new types of hybrid processes that took polymer science and engineering into the arena of functional materials. Its first piece of equipment, a hybrid platform, could take a liquid polymer, continuously spread it onto flat sheets, and dry or polymerize it into films. It also allowed incorporation of nano fibers. For example, it is used to prepare clear polymer films for liquid crystal displays. Additional research in the field involves transparent films that are also conductive, which is important for flexible electronic applications. Professor Frank Harris of the Department of Polymer Science was the University of Akron's principal leader in the proposal for Polymer Photonics and became the Photonics Team Leader for the consortium of partner universities.

Besides this laboratory, the University of Akron also planned to use Wright Center funds to construct a new fifty-thousand-square-foot laboratory near the Polymer Engineering Academic Center. The university would contribute $10 million to this construction—tentatively called the National Polymer Innovation Center—with an additional $3.5 million coming from other donors. It is scheduled for completion in 2009.[59] Proenza says the center will "take university research from the laboratory bench to the scale required for investment decisions."[60]

Acquiring funds for the building, however, is an ongoing effort. In September 2006, the University of Akron hired Birch, Horton, Bittner and Cherot to work on its behalf to secure federal funding and develop the National Polymer Innovation Center. Jon Devore, formerly a chief counsel and legislative director to Senator Lisa Murkowski (R-AK), is heading the university's lobbying efforts in Washington.[61]

The new Wright Center and local industry will have a close relationship. Goodrich, Goodyear, Honda, and Lockheed Martin will be responsible for suggesting ideas for future research in the hopes that it will result in useful commercial products, for instance, next-generation batteries, very thin polymer films for dashboard displays, improved polymer automobile bumpers, or artificial skin. Another important research area is photovoltaic materials to charge the batteries of a new airship, developed by Lockheed Martin, which could hover at sixty-five hundred feet above the earth. Clark Reed, the director of technical operations for Lockheed Martin Maritime Systems and Sensors, said, "The timing of this is absolutely phenomenal."[62] For the new airship to succeed, it will also require the use of strong, thin fibers composed of polymer nanocomposites, for the surface of the dirigible. The lighter the fibers, the more sophisticated and heavier the equipment used aboard can be.

Another major extension for the Center for Multifunctional Nanocomposites and Devices came in early 2007. Miko Cakmak was awarded an $8 million grant from the Ohio Third Frontier program to study "Functional Polymides and High Performance Nanocomposites." Polimides are polymers that possess properties that can withstand extreme electrical, thermal, and mechanical environments. These funds will integrate the efforts of the University of Akron, Kent State University, and the University of Dayton. The academic centers will also collaborate with Lockheed Martin, Kent Displays, General Electric Aviation, Maverick, Ethicon Endo, DuPont, GrafTech, and NanaSperse to develop products, including optical films for LCD displays, photovoltaic films, thermal management films for electronic devices, and ultraperformance 3D polimide nanocomposite parts for jet engines.

While important progress has been made, convincing Ohio legislators that polymers are vital to the future of the state is still an uphill battle. In November 2006, Proenza said he was still dismayed that "most people in state government still do not acknowledge that polymers are as large or larger than agriculture."[63] Proenza may be the "Next best hope to lead Akron toward finally achieving its long-held goal of building an economy truly worthy of the title 'Polymer Valley.' "[64] Along with Proenza's efforts, continued active and aggressive leadership in the College of Polymer Science and Polymer Engineering will play a central role in making this decades-old dream a reality.

POLYMER SCIENCE TODAY

Cutting-edge research in polymer science can involve unlikely and unusual creatures. One of them is the gecko lizard. "The scientific quest to make artificial gecko feet has taken a leap forward," wrote a *New York Times* reporter in 2005.[65] Unlocking this secret was important because it promised the creation of a strong adhesive that was glue free, reusable, and able to perform in the vacuum of outer space. In part, this was the result of the University of Akron's Ali Dhinojwala's work involving the reproduction of "van der Waals interactions" found on the feet of gecko lizards. These are weak molecular forces that bind solids and liquids together. Researchers are replicating the seemingly magical adhesive properties of the invisible hairs on the lizard's feet by growing microscopic carbon nanotubes in a laboratory. Applications are numerous, from robots that climb walls to mountain climbers who could scale rocks like Spider Man.[66] Dhinojwala has called his research "nano-velcro." Investigating natural polymers, Kelley said, represented the "cutting edge of research in materials."[67]

Gecko research is just one example of the important polymer research projects ongoing in Akron today. The Department of Polymer Science now boasts twenty active faculty and six emeritus professors. The faculty and students work together in twelve laboratories in the Goodyear Polymer Center maintained by the Maurice Morton Institute of Polymer Science. These include the Applied Polymer Research laboratory, a physical testing and analytical measurement laboratory providing technical service to industry. The Chemical Synthesis Laboratories include eighteen facilities, spread through nine floors of the

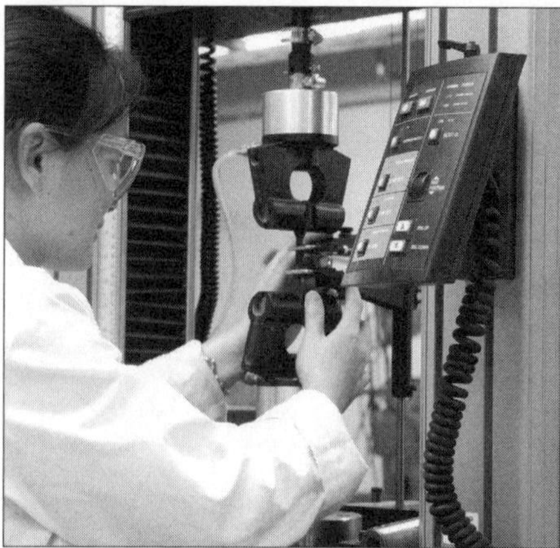

The Applied Polymer Research Center (APRC). Photos by John Ashley.

A polymer specimen mounted in an Instron tensile test machine to measure its tensile properties.

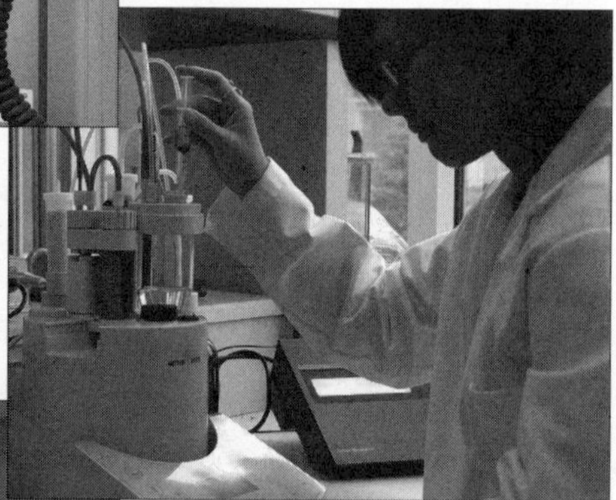

APRC has access to instruments that are maintained by APRC as well as those instruments that are available through the Institute of Polymer Science.

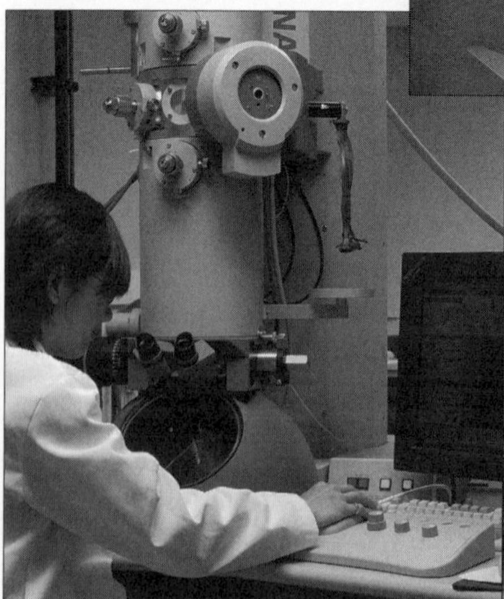

APRC has capabilities in Transmission and Scanning Electron microscopy as well as state of the art NMR instrumentation.

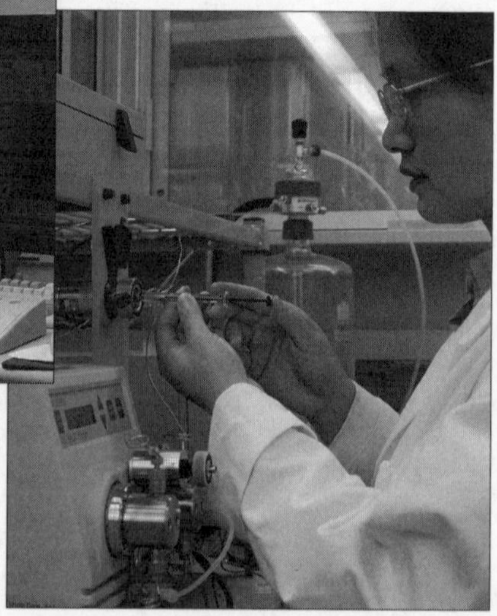

APRC maintains a full-time staff that is focused on contract type services for all polymer related industries.

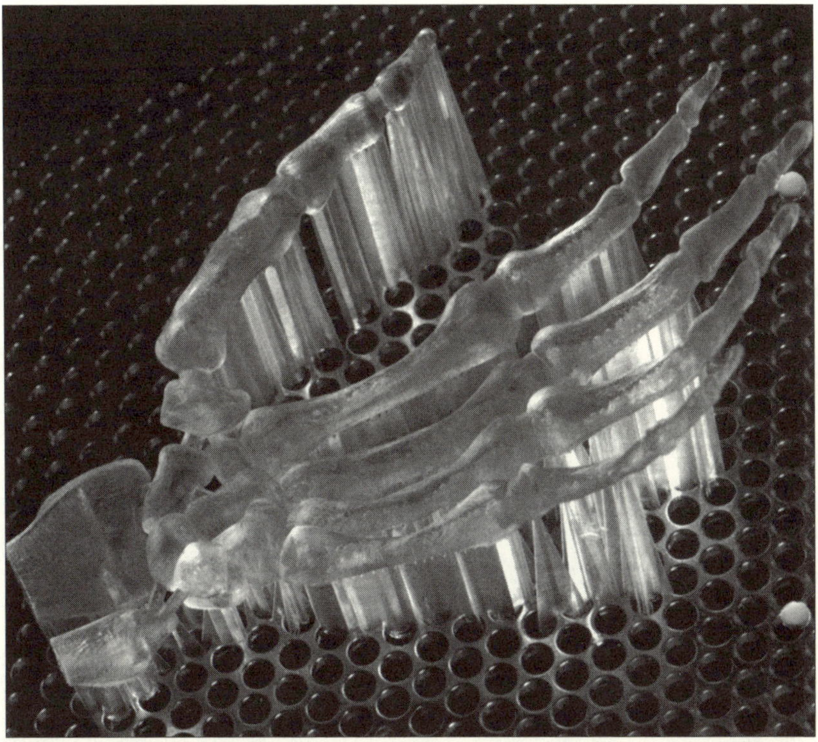

A state-of-the-art stereolithography machine created this model of a skeletal human hand. The Institute of Polymer Science acquired this equipment from the Ohio Board of Regents' Ohio Rapid Process Development Consortium. *Polymer* (Summer 1997).

Goodyear Polymer Center, supporting a broad spectrum of research, including the development of new materials and synthesis techniques; biocompatible materials in the areas of artificial veins, artificial arteries, and synthetic vertebrate pads; bandages that speed healing; controlled-release medicines; membranes that will dispense insulin for diabetes patients; and polymer materials for faster optical data transfer rates.

Other laboratories include a Clean Room used for the fabrication of components for liquid crystal displays in a dust-free environment. A Macromolecular Modeling Center is used for the development of computer modeling of polymers, enabling researchers to predict properties of new materials under development based on their molecular structures. Modeling on a computer versus actually carrying out the physical experiments in a traditional laboratory saves a tremendous amount of time. An electron microscopy laboratory uses transmission and scanning electron microscopes to greatly magnify very small objects. It focuses an electron beam contained within a vacuum to illuminate the object and projects its electromagnetic fields to magnify the image. The nuclear magnetic resonance (NMR) facility consists of four spectrometers used to determine chemical structures of polymers dissolved in a solution. Another solid state NMR can perform the same function on solid materials. The Physical Properties Laboratories make up thirty-two distinct laboratories in the Goodyear Polymer Center, which are used to test the characteristics of polymers, determine molecular structure, and analyze their mechanical properties. There are also support laboratories—an electronics shop and a machine shop, where technicians maintain and improve the variety of scientific research equipment used throughout the building, and a glassblowing facility.

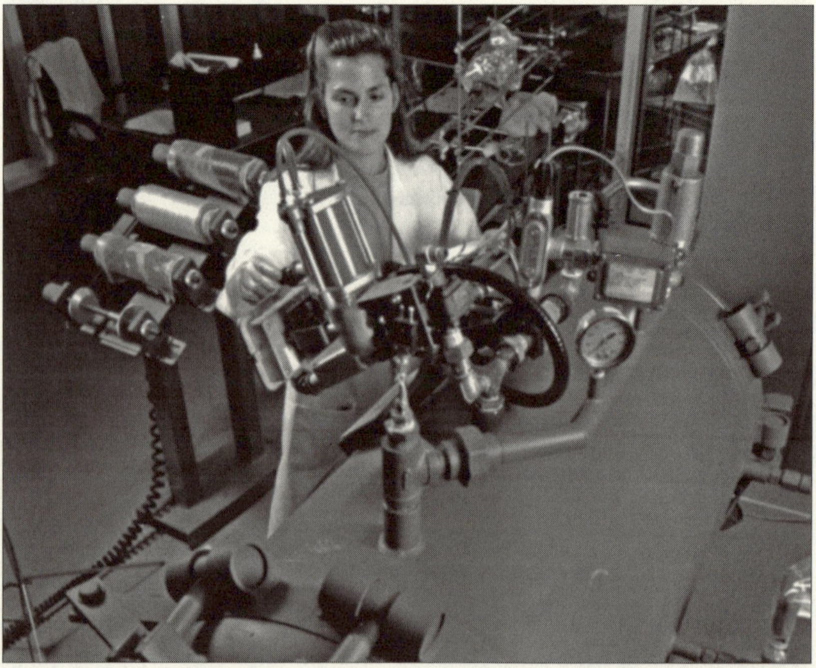

Tracey Richey, a polymer science Ph.D. student, examines the autoclave installed at the composites laboratory in the Goodyear Polymer Science building. The autoclave became part of Akron's "one-stop" laboratory where researchers could produce new polymers, assess their physical properties, produce carbon-fiber reinforced composites, and determine their properties. *Polymer* (Winter 1993).

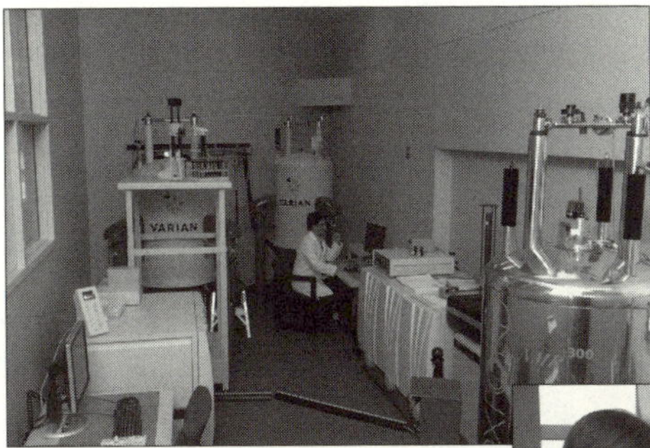

Polymer Science Today. Photos by John Ashley.

Nuclear magnetic resonance instruments for molecular characterization.

TGA (thermo gravimetric analysis) and DSC (differential scanning calorimetry) are two thermal research instruments.

Simultaneous wide and small angle x-ray scattering instrument.

Field emission scanning electron microscope.

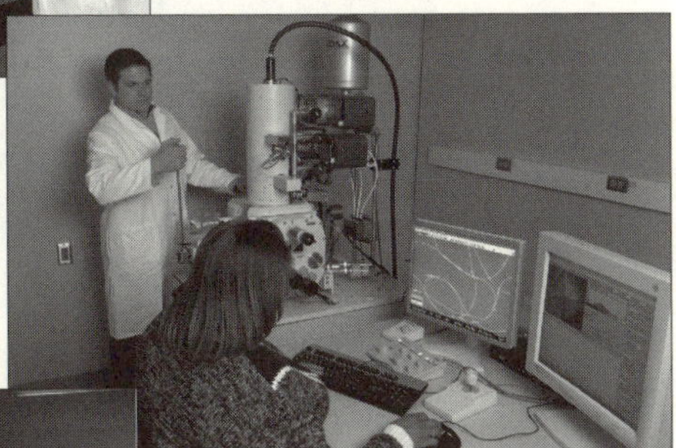

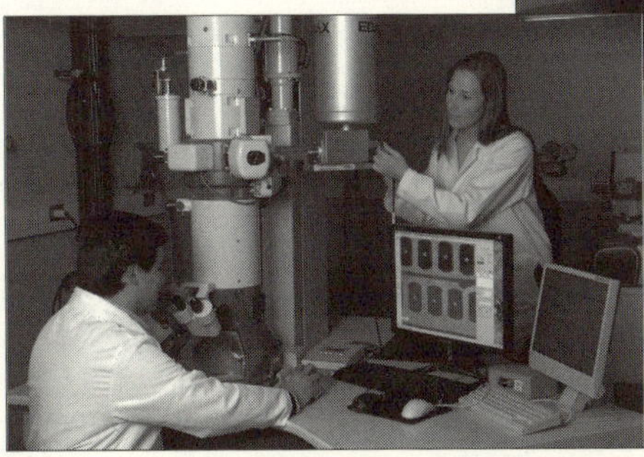

High performance transmission electron microscope.

(*left*) Advanced computer modeling of polymeric materials is done using state-of-the-art computer hardware and software available in Gustavo A. Carri's laboratories. *Photo by John Ashley.*

(*below*) Graduate students setting up light scattering experiments on polymeric nanostructures in Alexi Sokolov's laboratory. Light helps to probe mechanical properties of structures invisible in the best optical microscopes due to their extremely small size. *Photo by John Ashley.*

Two final facilities include the Solution Characterization Laboratory and the Surface Spectroscopy Laboratory.

While scientific research and the lure of commercial development is a central activity of the polymer faculty and its graduate students, teaching remains part of the essential pedagogical mission of the College of Polymer Science and Polymer Engineering. The teaching mission is never far from the heart of the polymer program because from these educational activities will emerge the polymer leaders of tomorrow. Historically, the pioneering rubber chemistry course offered by Charles Knight in 1909 was the starting point of the university's current international reputation in the field. Although undergraduate students were the traditional emphasis of the university in the days of Knight, today's emphasis is a comprehensive core of master's and doctoral programs. The Goodyear Polymer Center houses teaching laboratories specifically designed for instructing graduate students in experimental techniques in polymer research. Primarily, the college focuses on graduate students, while offering undergraduate polymer courses for other science and engineering departments within the university. The department recognizes individual student excellence in its program through a variety of annual awards.[68]

Many of the students who have passed through the program have become distinguished alumni, serving as leaders in the academic and industrial communities. While it is impossible to address the significance of all of its graduate students, the Department of Polymer Science has recently instituted a Distinguished Alumni Award that has honored eight individuals since 1997. James E. McGrath (Ph.D., polymer chemistry, 1967) spent most of his career at Virginia Tech, where, in 1989, he took a position as director of the NSF Science and Technology Center at the university.[69] Lewis Fetters (Ph.D., polymer chemistry, 1962) was a professor of polymer science at the University of Akron from 1967 to 1983. Nissim Calderon (Ph.D., polymer chemistry, 1962) worked at Goodyear after graduation, and in 1964 he discovered olefin metathesis, which appears today in over three hundred patented applications for petrochemical products, perfumes, and polymers. In 1986, he became the vice president of research at Goodyear, where he oversaw over four hundred scientists and engineers.[70] Russell A. Livigni (Ph.D., polymer chemistry, 1960) had a thirty-year career at GenCorp, where he put his name on over thirty patents, became the vice president and director of research, managed a $36 million annual research and development budget, and oversaw a group of one hundred people.[71] Frank Kelley (Ph.D., polymer chemistry, 1961) won the distinguished alumni award in 2003. Karl G. Kempf (Ph.D., polymer science, 1977) won an Academy Award for his work on special effects in the Superman movies, worked on automation for the original Space Station at McDonnell Douglas, and in 1987 joined Intel, where he is now an Intel Fellow and Director of Decision Technologies. Sharell Mikesell

(*above*) Russell A. Livigni. *Polymer* (Fall 1990).
(*right*) Nissim Calderon. *Polymer* (Spring 1990).

(Ph.D., polymer chemistry, 1971) became vice president for technology at Owens-Corning Fiberglass. In 1995, he left Owens, serving as vice president of technology at Advanced Elastomer Systems for six years until he became the first executive director of the Ohio Polymer Strategy Council in 2001.[72] Robson Storey (Ph.D., polymer science, 1983) has spent twenty-three years at the University of Southern Mississippi, where he is today a professor of polymer science and engineering. These few distinguished polymer science alumni—McGrath, Fetters, Calderon, Livigni, Kelley, Kempf, Mikesell, and Storey—are just a small sample of more than one thousand graduates whose influence has been felt throughout the world.

Alongside the traditional emphasis on graduate education, today, the program's leaders continually look for new and innovative ways to teach a variety of students about polymers. This includes community programs like Upward Bound, programs for gifted high school students, and opportunities to demonstrate scientific experiments for elementary and middle school children. Dean Frank Kelley has also taught a course with Ray Gehani, a professor from the College of Business, called Management Decisions in the Polymer Industry; it was targeted to MBA students who wanted to become more aware of the polymer industry.[73]

It might seem unusual for a university to be interested in high school students, but younger students represented the future. Mark Foster, chair of the Department of Polymer Science, said, "Probably the number one challenge that we face is a growing ambivalence of American students to the sciences, particularly in high school."[74] One reason there are so few Americans pursuing scientific careers in graduate schools was growing disinterest in these subjects in high school.[75] As a result, the polymer program began to look for opportunities to foster interest in science and inspire a younger generation to think about polymers. In 1992, two local high schools, in Hudson and Stow, linked their computers to the University of Akron's polymer laboratory. This enabled these young students to use advanced programs that, for example, showed rotating three-dimensional images of polymer molecules. The students could then manipulate the images by adding other molecules to see how they would react. James Parker, a biology teacher at Stow, along with Kelley and Wayne Mattice, a polymer science professor, headed up the program. Parker said, "the potential is tremendous."[76]

These initiatives continued and were enhanced through the federally funded Upward Bound program, which exposes high school students throughout the region to math and science through polymers. Officially called the "Upward Bound

Mark Foster. *Polymer* (Fall 1995).

Regional Math/Science Program," it is jointly controlled by the University of Akron and the U.S. Department of Education, and offers a six-week residential session in the summer for fifty-eight high school students from area schools, including Buchtel, East, Ellet, Firestone, Garfield, Kenmore, and North High Schools. Students are exposed to experiments, do an independent research project with university professors, hear guest lecturers, receive career counseling, and take field trips to businesses such as Advanced Elastomer Systems.

There is also a polymer educational mission to begin instruction for even younger students. One example was at King Elementary School in Akron, where a program was begun for children to learn about polymers in the early grades. In 1992, it became part of a pilot project

In 1992, fifth grade students at King Elementary School became the pilot school in a partnership between the Akron Public School system and the College of Polymer Science and Polymer Engineering. Here two students use the beads to understand polymers linked in a long chain. *Polymer* (Spring 1992).

called "Hands-On Science" between the Akron Public Schools and the College of Polymer Science and Polymer Engineering. Doctoral candidates in polymers visited biweekly, offering to give enthusiastic lessons involving polymers, chemistry, and mathematics.[77] King School has interacted most recently with the university through the Global Polymer Academy. In 2006, one of its science teachers held a "Family Science Day," and representatives from the polymer program gave presentations. The Akron Global Polymer Academy also offers third grade lesson plans and teaching suggestions on its web site.

Polymer training is also necessary in continuing education for workers already in the plastics and rubber industry. In fall 1993, the university established the Akron Polymer Training Center in an 18,500-square-foot teaching facility on its campus. It began serving the region's academic and industrial needs by offering a variety of credit and non-credit courses. Nancy Clem, the center's initial director, said, "From the outset, our curriculum planning has been centered on meeting the specific needs of people involved in plastics and rubber processing. . . . As a result, the response has been terrific."[78] To accommodate those with full-time jobs, many of the courses are offered in two-hour blocks in the evening. The training center itself contains three classrooms, two polymer processing laboratories, and a laboratory devoted to chemical measurement and instrumentation. It offers plastics courses (a twin screw extrusion seminar, color technologies, complete injection molding, and plastics recycling) and rubber courses (chemistry and physics of elastomers, compounding for performance, design of experiments, injection molding of elastomers, mixing and processing of rubber, rubber technician training, solving problems in rubber compounding and processing, and testing of rubber processability and dynamic properties). After four years of operation, it had trained over seventeen hundred students and continues as a valuable resource today, under the direction of Tayba L. Tahir.

Kelley believed that promoting polymers to a diverse student base was an essential component of the college's pedagogical mission. While graduate education and research remained its central activity, he saw the outreach efforts spreading the university's impact across a much broader spectrum of the population. Kelley said that, as a

Advanced Injection Molding course taught at the Akron Polymer Training Center. The VanDorn Demag 170 injection molder, is one of two of these machines at the training facility. *Photo courtesy of the Akron Polymer Training Center.*

result of all of these programs, "We have taken significant steps to enhance the University of Akron as a major center of polymer education at every level."[79]

A HALF CENTURY OF POLYMER PH.D.'S

In August 2006, the University of Akron approached a milestone. It had been fifty years since it had begun one of the first Ph.D. programs in polymer chemistry in the world. Since the first five doctoral candidates graduated in 1959, the university has gone on to develop twenty-three other Ph.D. programs; it has also graduated over one thousand Ph.D.'s in polymer science and polymer engineering. Today the College of Polymer Science and Polymer Engineering has thirty-four full-time faculty, two hundred graduate students, sixty postdoctoral fellows, and four main buildings on campus. Funding for the college reaches about $10 million per year, both from industry and government. Former department chair Rod Quirk, who led a committee to mark this occasion, said, "I think it's one of the most important milestones in the university's history."[80]

The celebration drew attendees from all over the world. Speaker Arden Bement, director of the National Science Foundation, discussed the climate for scientific research in the United States today. The era of long-term basic research is changing significantly, he said, explaining "We're finding out the long term is getting shorter all the time." He discussed the "more demanding pace" that stemmed from a world expecting more immediate results, which becomes increasingly connected every day. "We are confronted with the globalization of science and engineering."[81]

The University of Akron had clearly been impacted by the forces of change that Bement described. Its polymer programs were a representation of the world's interest in polymer studies, with faculty and students from nearly every part of the globe coming to Akron. Its programs have also been affected by the fast pace of change, most evidently with the termination of EPIC and OPED. Both of these programs lost state support because they had not created new polymer products or polymer start-up companies fast enough. The proponents of both organizations argued that if they had more time to achieve their goals, they would have been successful. Bement concluded, "Should we be worried? Perhaps. Should we panic? Definitely not."[82]

The fiftieth anniversary also marked another milestone, as several months before the celebration Frank Kelley announced he would be stepping down from his position as dean and retiring. He had served as dean for eighteen years and had spent a total of twenty-eight years at the university, after returning to campus in 1978 as the director of the Institute of Polymer Science. He had been a member of one of the first polymer chemistry Ph.D. classes at the university, and he believed his greatest legacy was his own graduate students, now in leadership positions in academia and industry.[83] Another key legacy was a strong faculty with stable funding ensured for the future through established endowments. The physical facilities erected during his tenure—the Goodyear Polymer Center and the Polymer Engineering Academic Center—will always remain as a visible reminder of his presence in Akron.

Professor Joseph Kennedy perhaps best captured the significance of the Kelley years in a recent interview. Kennedy identified Kelley as the "motor" of the entire polymer program. He attributed the formation of the College of Polymer Science and Polymer Engineering to Kelley's foresight and influence, and he believed this was unprecedented, perhaps in the entire world. Kennedy said Kelley "had the vision to put this together. He also had the administrative capabilities to manage, develop, and nurture young people and bring them up to the international level." He added, "We are beholden to him. Without him, there would be no University of Akron polymer program."[84]

In December 2006, the University of Akron hosted another large gathering to commemorate the end of the Kelley tenure.[85] Several hundred of his former students, faculty, and friends celebrated a career that had been important to the university, the polymer industry, and the city. A highlight was a surprise given to Kelley by his daughter, Katie Bruno, and President Luis Proenza. Fourteen years earlier, Kelley had the unique privilege of awarding his daughter a Ph.D. in polymer science from the University of Akron. At the time, he said, "I got to put the hood over her head—how many fathers get to do that?"[86] This time, it was his daughter's turn to give her father a Ph.D.—an honorary doctorate of humane letters—and she placed the hood over his shoulders in an emotional culmination to her father's professional career at the university.

One of the most unique pieces of the Kelley legacy is a forty-foot glacial-blue crystal polymer tower that stands outside the Goodyear Polymer Center. It was Kelley's idea to commission a dramatic piece of artwork to symbolize the polymer pioneering activities at the university. Renowned glass artist Dale Chihuly, who began working with polymers in 2000 as a revolutionary new material, agreed to develop a project for the university. Part of the deal was that Chihuly could tap the polymer expertise of its scientists and engineers to teach him more about this unique material. Fundraising took six years, and the total cost of the sculpture was $475,000. Bayer Materials Sciences donated the polymers for the project, the Ohio Arts Council contributed the concrete base, and Mary Myers (a benefactor of the university and the Akron Art Museum) and A. Schulman (a Fairlawn, Ohio, plastics

The Dale Chihuly polymer sculpture that stands in front of the Goodyear Polymer Center today. *College of Polymer Science and Polymer Engineering Archives.*

compounder) gave additional funding. Chihuly constructed his polymer masterpiece out of eighty blue Polyvitro crystals, each weighing between sixty and eighty pounds. Some speculated the piece was an important symbolic representation for the region. *Akron Beacon Journal* art critic Dorothy Shinn wrote, "In Akron the tower represents not only the UA Goodyear Polymer Center, but a large chunk of the economy of Akron, Summit County, and the state of Ohio."[87] President Luis Proenza called it an "artistic expression of the science that has distinguished the University of Akron for so long."[88]

Reflecting back and looking forward, Kelley believes the polymer programs at the University of Akron are at a critical moment. He said, "I think we are at a watershed time, and this program can either sink to the depths or rise to the heights. It all depends on acquiring and retaining the quality of faculty we have. The faculty bring the reputation; they bring the funds for doing the job and they attract the students."[89] In a large measure, this future will be guided by the new dean, Stephen Z.D. Cheng, selected in August 2007 to assume the leadership of the College of Polymer Science and Engineering. He arrived at the university in 1987, and his twenty years of experience in Akron

Stephen Z.D. Cheng (*right*), the new Dean of the College of Polymer Science and Polymer Engineering, with Matt Graham, his former Ph.D. student (*left*). *From the personal collection of Stephen Z.D. Cheng.*

have given him a unique understanding of its past, as he now will guide it into the future. Embarking upon this new challenge, Cheng wrote:

> From its inception over fifty years ago, the University of Akron polymer program has been at the forefront in developing new exotic and everyday materials that have raised peoples' standard of living world-wide. Under Dr. Frank N. Kelley's leadership, the College of Polymer Science and Polymer Engineering extended its impact beyond just the academic sphere into supporting industry and improving the community. As the new Dean, I am grateful for the opportunity to realize our new vision of increasing the scientific, industrial, and societal impact of the College of Polymer Science and Polymer Engineering. The field of polymers is broadening into exciting emerging areas including information and bio-medical technology. As the largest polymer institution in the U.S., we will actively seek to advance these

> emerging areas through the development of innovative science and technology. By communicating new understandings and translating these findings into new applications, the College of Polymer Science and Polymer Engineering will continue to contribute to the advancement of society as a whole.[90]

Cheng comes to a program with a long history, a century in length, adding a new name to the short roster of leaders of the rubber chemistry and polymer programs in Akron's history—Charles Knight, Hezzleton Simmons, G. Stafford Whitby, Maurice Morton, and Frank Kelley. Cheng will bear a heavy burden of past expectations, but he will also benefit from the strong foundation these men laid. He will ensure that this critical moment becomes an opportunity to protect, cultivate, and promote the program's heritage for the future.

CONCLUSION—AKRON AND THE POLYMER AGE

In February 2000, a chemist writing in the journal *Chemistry & Industry* suggested that the emergence of polymers has resulted in a new material age for civilization. He said, "The importance of this material is such that our present day society may well be remembered as 'the Polymer Age,' just as we refer to the Iron and Bronze Ages."[91] Institutions—academic, industrial, and governmental—have been the driving forces that have enabled this Polymer Age. Of the academic institutions that have played a role in its emergence, the University of Akron unquestionably stands among the elite.

In 1996, *U.S. News and World Report* for the first time ranked the top five polymer university programs, separate from chemistry departments.[92] The results of the inaugural ranking demonstrated northeastern Ohio's strength in polymers. The University of Massachusetts at Amherst earned first place, followed by the University of Akron in second. Tied for third were the California Institute of Technology and the University of Southern Mississippi. In fifth place was Case Western Reserve University. While it was an honor to be placed higher than every other academic institution but one, the University of Akron has remained at this position for the last decade. Luis Proenza does not consider second place or a silver medal a victory. He said, "We don't want to be second to anybody. It's important, given how close we are, to become number one."[93] He believes that overcoming the stigma of second place will strengthen the polymer program itself and also bring important contributions to the regional economy. This is Proenza's polymer goal for the future.

Despite Proenza's aspirations, the University of Akron's polymer program has already made tremendous contributions not only to the academic pursuits of polymer science and polymer engineering, but also to the local economy itself. In 2006, the editor of the *Akron Beacon Journal* reflected on the importance of the polymer program at the University of Akron in the wake of the decimation of the rubber industry. He said, "The value of the UA polymer college goes beyond jobs and dollars. Its presence has helped define the character of the community."[94] It does this through the elusive and sometimes unquantifiable measurement of "innovation." The dreams of a Polymer Valley may not have reached the expectations of its founders or the surrounding community, but the polymer program still has great worth to the region. Samir Gibara, the chairman of Goodyear, has said the city is better off with a world-class polymer program than it would be with two more rubber plants: "I mean, everybody has plants. You want a competitive advantage. To be distinctive." From his perspective, it is not just the university, but the city of Akron that has become distinctive because of this polymer program. It is the academic phoenix that has helped rejuvenate the city in which it lives. Gibara concluded that the College of Polymer Science and Polymer Engineering is the "intellectual legacy of the tire industry to the city."[95]

This legacy has been built over nearly a century of intellectual activity which has defined and shaped a discipline that began as rubber chemistry and evolved into polymer science and polymer engineering. Among its highlights: Charles Knight offering the world's first course in rubber chemistry in 1909; Hezzleton Simmons and G. Stafford Whitby directing vital scientific and administrative elements of the synthetic rubber program during World War II; Maurice Morton and Norman Auburn establishing the first Ph.D. program in polymer chemistry in 1956 and working together to help build a world-renowned Department and Institute of Polymer Science in the 1960s. Frank Kelley elevated the program to serve as a focal point for economic renaissance in Akron in the 1980s and built unprecedented new physical facilities in 1991 and 2001. James L. White established a new program in polymer engineering in the 1990s and contributed mightily to the formation of a new discipline; Luis Proenza and leaders in the polymer community continue to work to reinforce and strengthen Ohio's claim as the polymer state at the start of the twenty-first century.

No other academic institution in the United States, or the world, has had such a long and essential influence upon the disciplines of rubber chemistry, polymer science, and polymer engineering. This unbroken chain of excellence, starting with Charles Knight at the turn of the twentieth century and extending to the blossoming of an entire polymer college today, offers opportunities for future generations. As a society, our mastery of polymer molecules will continue to yield unimagined material possibilities. We are all living in the Polymer Age.

APPENDIX A

UNIVERSITY OF AKRON CHEMISTRY GRADUATES IN THE RUBBER INDUSTRY, 1910–1941

The following table includes a selection of University of Akron chemistry graduates from 1910–1941 and lists positions they later held in the rubber industry:

NAME	CLASS	POSITION	COMPANY	CITY
Belden, Russell D.	1910	Vice President	Dunlop Tire & Rubber	Buffalo, NY
Theiss, Fred C.	1910	Rubber Compounder	Sun Rubber	Akron, OH
Arbogast, Harry L.	1912	Technical Superintendent	West American Rubber	Altadena, CA
Grafton, Dr. Elvah H.	1913	Manager, Control and Development	Tompkins Rubber	Secane, PA
Jackson, Lee R.	1913	President	Firestone Tire & Rubber	Akron, OH
Sidnell, Albert E.	1914	Vice President	Seiberling Rubber	Akron, OH
Wilson, Robert F.	1914	Technical Manager	Firestone Tire & Rubber	San Gabriel, CA
Hanna, Ezra L.	1915	Director of Laboratories	Davol Rubber	Scituate, RI
Crisp, Park P.	1916	President	Firestone Industrial Products	Akron, OH
Frick, Carl E.	1916	Production Manager	Van Cleef Bros. Rubber	Chicago, IL
Yakee, Bert H.	1916	Chief Chemist	Goodyear Tire & Rubber	Gadsden, AL
Carlton, Clinton	1917	Technical Director, Rubber	J.M. Huber	Borger, TX
Holloway, Harold S.	1919	Chemist	Firestone Tire & Rubber	Akron, OH
Purdy, Walter	1919	Chemist	Shanco Plastics & Chemical	Tonawanda, NY
Endres, Herbert A.	1921	Assistant Manager, Research	Goodyear Tire & Rubber	Akron, OH

NAME	CLASS	POSITION	COMPANY	CITY
Snyder, Harold L.	1921	Rubber Tech	Griffen, Cooper, Keck	Vallejo, CA
Barnes, Garrett H., Jr.	1922	Manager, Airfoam Division	Goodyear Tire & Rubber	Akron, OH
Christensen, Chester W.	1922	General Sales Manager	Monsanto Chemical	Akron, OH
Eckert, Herman K.	1922	Plant Manager	Monsanto Chemical	Texas City, TX
Hardman, Albert F. Sr.	1923	Research Chemist	Goodyear Tire & Rubber	Akron, OH
Penfield, Harold W.	1923	Rubber Chemist	Birdsboro Steel Foundry	Reading, PA
Ruch, Elden	1923	Chief Compounder	Firestone Tire & Rubber	Memphis, TN
Schrank, Harry P.	1924	Vice President	Seiberling Rubber	Akron, OH
Baker, Deming E.	1925	Laboratory Manager	Monsanto	Akron, OH
Carman, Frank H.	1925	Secretary	Plastics Materials Manufacturers Assoc.	Washington, D.C.
Thorn, Homer Z.	1925	Chemist	Firestone Industrial Products	Swansea, MA
Mitchell, Peter B.	1926	Development	Firestone Tire & Rubber	Akron, OH
Kilbourne, Dr. F.L. Jr.	1927	Chief Chemist	Xylos Rubber	Akron, OH
Wagstaff, Donald A.	1927	Chemist	Firestone Tire & Rubber	Akron, OH
Balazs, Gabriel G.	1928	Technical Director	Goodyear Tire & Rubber	Akron, OH
Brodt, Carl W.	1928	Manager, Development	Firestone Industrial Products	Swansea, MA
Cole, Dr. Otis D.	1928	Assistant Director, Research Labs	Firestone Tire & Rubber	Akron, OH
Crankshaw, Jesse	1928	Chemist	Midwest Rubber Reclaiming	Akron, OH
Finney, Addis	1929	Factory Manager	Firestone Tire & Rubber	Essel, Switzerland
Nicol, Sheldon C.	1929	Compounder	Goodyear Tire & Rubber	Cuyahoga Falls, OH
Baker, Lawrence M.	1930	Manager, Product Compounding	General Tire & Rubber	Akron, OH
Brueggeman, Elmer C.	1930	Chief Chemist	Seiberling Latex	Akron, OH
Finley, Milton	1930	Development Manager	Goodyear—Lincoln Division	Lincoln, NE
Galloway, F. Merrill	1930	Technical Superintendent	Quaker Rubber	Langhorn, PA
Lenhart, George B.	1930	Chief Chemist	Wilson Rubber	Canton, OH
Dete, Leo J.	1931	Chief Chemist	Carlisle Tire & Rubber	Carlisle, PA
Dobberstein, Arthur W.	1931	Vice President	Industrial Latex	Dover, DE
Peabody, Donald W.	1931	Research Chemist	Goodyear Tire & Rubber	Akron, OH

NAME	CLASS	POSITION	COMPANY	CITY
Hohl, Dr. Willy	1932	Manager, Rubber Department	Suhner Wire & Rubber	Herisau, Switzerland
Brandau, Karl	1933	Rubber Chemist	Firestone Tire & Rubber	Akron, OH
D'lanni, Dr. James D.	1934	Section Head, Synthetic Rubber	Goodyear Research Lab	Akron, OH
Tausch, Harry W.	1934	Airfoam Development	Goodyear Tire & Rubber	Akron, OH
Ambelang, Dr. Joseph C.	1935	Research Chemist	Firestone Tire & Rubber	Akron, OH
Bosshard, Alfred	1935	Chemist	Firestone Tire & Rubber	Akron, OH
Boyd, Dr. Thomas	1936	Plastics Research	Monsanto Chemical	Springfield, MA
Carter, Donald E.	1936	Head, Thermo-plastics Laboratory	Bakelite Division, UCC	Rahway, NJ
Forster, William E.	1936	Manager	Goodyear Plant	Argentina
Marcelo, Jose P.	1936	President	Marcelo Rubber	Quezon City, Philippines
Cashion, Clifton G.	1937	Technical Sales	Xylos Rubber	Akron, OH
Laatsch, Herman	1938	Technical Superintendent	Goodyear Tire & Rubber	Dartmouth, MA
Albert, Dr. Harry E.	1939	Research Chemist	Firestone Tire & Rubber	Akron, OH
Glazer, Emmett J.	1939	Research Chemist	Goodyear Tire & Rubber	Akron, OH
Gray, Russell Houston	1940	Senior Chemist (Research)	E.I. DuPont de Nemours	Rochester, NY
Hausch, Dr. Walter R.	1940	Research Chemist	Firestone Tire & Rubber	Akron, OH
Hoertz, J. Bruce	1940	Research Chemist	B.F. Goodrich	Akron, OH
Hollis, Arthur L.	1940	Technical Manager	B.F. Goodrich	Akron, OH
Klar, Kenneth K.	1940	Chief Chemist	Industrial De Pneumatico Firestone	Sao Paulo, Brazil
Semgen, Stephen T.	1940	Research Chemist	B.F. Goodrich	Akron, OH
Snyder, Joseph E.	1940	Rubber Analyst	Brush Beryllium	Akron, OH
Becker, Elwyn H.	1941	Chemist	Wilson Rubber	Akron, OH
Dean, David J.	1941	Chemist	Owen Rubber & Plastic	Akron, OH
Hershey, Robert M.	1941	Engineer, Plastics and Rubber	Hoover	Canton, OH
Krivitsky, Albin S.	1941	Chemist	Goodyear Tire & Rubber	Akron, OH
Loulan, James A. Jr.	1941	Comopunder	Goodyear Tire & Rubber	Gadsden, AL
Strobel, Everett H.	1941	Production Manager	Firestone Tire & Rubber	Akron, OH
Wahn, Harvey L.	1941	Chief Chemist	Hawkeye Rubber Manufacturing	Cedar Rapids, IA

APPENDIX B

SECRET AKRON WAR PRODUCTION BOARD MEETING, 1942

Below are attendees of the December 1942 "Secret and Confidential, War Production Board" meeting held in Akron's Mayflower Hotel.[1] They decided that the University of Akron would play a key administrative role in the United States synthetic rubber effort.

Baker, W.O.	Bell Laboratories
Bebb, R.L.	Firestone Tire & Rubber
Biggs, B.S.	Bell Laboratories
Borders, A.M.	Goodyear Tire & Rubber
Carr, Edward	Firestone Tire & Rubber
Clifford, A.M.	Goodyear Tire & Rubber
Crawford, R.A.	B.F. Goodrich
D'Ianni, J.D.	Goodyear Tire & Rubber
Dillon, J.H.	Firestone Tire & Rubber
Dunbrook, R.F.	Firestone Tire & Rubber
Ewart, R.H.	United States Rubber
Flory, P.J.	Esso Laboratories
Frolich, Per K.	Esso Laboratories
Fryling, C.F.	B.F. Goodrich
Fuller, C.S.	Office of Rubber Director
Garvey, B.S.	B.F. Goodrich
Gehman, S.D.	Goodyear Tire & Rubber
Harkins, W.D.	University of Chicago
Haworth, J.B.	Esso Laboratories
Howland, L.H.	U.S. Rubber

Japs, A.B.	B.F. Goodrich
Johnson, B.L.	Firestone Tire & Rubber
Juve, A.E.	B.F. Goodrich
Kemp, A.R.	Bell Laboratories
Kolthoff, I.M.	University of Minnesota
Marvel, C.S.	University of Illinois
Osterhof, H.J.	Goodyear Tire & Rubber
Prutton, C.F.	Case School of Applied Science
Semon, W.L.	B.F. Goodrich
Shepard, M.G.	United States Rubber
Shepard, N.A.	Office of Rubber Director
Sebrell, L.B.	Goodyear Tire & Rubber
Street, J.N.	Firestone Tire & Rubber
Stewart, W.D.	B.F. Goodrich
Swaney, M.W.	Esso Laboratories
Taylor, H.S.	Princeton University
Tobolsky, Arthur	Princeton University
Walton, C.W.	Goodyear Tire & Rubber
Whitby, G.S.	University of Akron
Wiegand, William B.	Columbian Carbon
Williams, R.R.	Office of Rubber Director

APPENDIX C

RUBBER AND POLYMER COURSEWORK: 1917, 1956, 2006

Over the past century of teaching rubber chemistry, polymer science, and polymer engineering, the instruction has evolved from a single course to an entire college. A selection of educational offerings in 1917, 1956, and 2006, demonstrate this evolution. In 1917, Hezzleton Simmons published the lectures to his rubber chemistry course in *The Rubber Age*. At the time, it was the only such academic course of its kind in the United States. The titles of those lectures are as follows:[2]

LECTURE #	LECTURE TITLE
Intro	The Laboratories and Equipment of the University of Akron
1	The History of Caoutchouc
2	Rubber of the Amazon Basin
3	African Rubbers Including Those from Madagascar
4	Central American Rubbers
5	Rubber Plantations and Their Development
6	Discussion of Colloids
7	The Colloidal Action of Crude Rubber and Its Applicability in Rubber Manufacture
8	Different Means of Coagulation
9	Theory of the Constitution of Rubber
10	Synthetic Caoutchouc
11	Chemical and Physical Testing of Crude Rubber
12	The Manufacture and Use of Inorganic FIllers

13 The Manufacture and Use of Organic Accelerators
14 The Manufacture and Use of Rubber Substitutes
15 Theories of Vulcanization
16 Methods of Reclaiming Rubber
17 Preparation of Crude Rubber for Manufacturing
18 The Principles of Compounding
19 Chemical Analysis of Manufactured Rubber
20 Physical Testing of Compounded Samples

The following list identifies the complete number of polymer academic courses offered throughout the United States in 1956, the year that the University of Akron began its Ph.D. in polymer chemistry.[3]

SCHOOL	# OF COURSES	COURSES OFFERED
University of Akron	9	Chemistry of Polymers, Chemistry of Polymers Laboratory, Chemistry of Latex Laboratory, Chemistry of Rubber Technology, Chemistry of Rubber Laboratory, Physical Chemistry of High Polymers, Experimental Physical Chemistry of Polymers, Mechanical Behavior of Polymers, Research in Rubber and Plastics
Case Institute of Technology	7	Technology of Plastics, Physical Chemistry of Polymerization, Physical Chemistry of High Polymers, Mechanical and Electrical Properties of High Polymers, Organic Chemistry of High Polymers, Advanced Technology of High Polymers, High Polymer Laboratory
University of Southern California	6	Colloid Chemistry of High Polymeric Materials, Organic Chemistry, Basic Rubber Technology, Rubber Technology II, Rubber Technology III, Rubber Technology Laboratory I and II
Michigan State University	5	Polymerization, Plastics, Surface Coatings, Technology of Fibers and Plastics, Surface Chemistry
Polytechnic Institute of Brooklyn	4	Introductory Polymer Chemistry, Organic Polymer Chemistry, Polymerization Kinetics, Mechanical Behavior of High Polymers

SCHOOL	# OF COURSES	COURSES OFFERED
University of Michigan	4	Introduction to High Polymers, Polymerization Process, Structure and Properties of High Polymers, Organic Chemistry of Synthetic Polymers
Wayne University	4	Technology of High Polymers, Principles of Chemical Technology, Physical Chemistry of Polymers, Seminar in Polymer and Colloid Chemistry
Cornell University	3	Technology of High Polymers, Organic Chemistry of High Polymers, Physical Chemistry of High Polymers
Illinois Institute of Technology	3	Principles of Rubber Compounding I, Principles of Rubber Compounding II, Colloidal and Amorphous Materials
Massachusetts Institute of Technology	2	Colloid Science of High Molecular Compounds, Colloid Science of Silicon
Northeastern University	2	Rubber Technology I, Rubber Technology II
University of Rhode Island	2	Rubber Compounding, Rubber Chemistry
Gidley Research Institute	1	Rubber Chemistry and Technology
John Marshall Law School	1	Basic Course in Rubber Compounding
Kent State University	1	Industrial Chemistry
New York Rubber Group	1	Elastomer Technology
Princeton University	1	Plastics Program
University of Buffalo	1	Elements of Polymer Chemistry
Villanova University	1	Basic Rubber Technology

Below are the graduate-level courses offered in 2006 at the College of Polymer Science and Polymer Engineering, a half-century after the founding of the polymer chemistry Ph.D. program.

Core graduate-level courses for the master's and doctoral degrees in polymer science, 2006:

Master's Degree Core

9871:601 Polymer Concepts

9871:613 Polymer Science Laboratory

9871:631 Physical Properties of Polymers I

9871:674 Polymer Structure and Characterization
9871:701 Polymer Technology

Ph.D. Degree Core

9871:601 Polymer Concepts
9871:602 Synthesis and Chemical Behavior of Polymers
9871:704 Condensation Polymerization
9871:705 Free Radical Reactions in Polymer Science
9871:706 Ionic and Monomer Insertion Reactions
9871:674 Polymer Structure and Characterization
9871:675 Polymer Thermodynamics
9871:631 Physical Properties of Polymers I
9871:632 Physical Properties of Polymers II
9871:701 Polymer Technology I
9871:702 Polymer Technology II
9871:703 Polymer Technology III
9871:613 Polymer Science Laboratory
9871:607 Polymer Science Seminar I
9871:608 Polymer Science Seminar II

Core graduate level courses for the master's and doctoral degrees in polymer engineering, 2006:

9841:611 Structural characterization of polymers with electromagnetic radiation
9841:621 Rheology of polymeric fluids
9841:622 Analysis and design of polymer processing operations I
9841:631 Engineering properties of solid polymers
9841:641 Polymeric materials engineering science
9841:601 Polymer engineering seminar
9841:623 Analysis and design of polymer processing operations II
9841:642 Engineering aspects of polymer colloids
9841:650 Introduction to polymer engineering
9841:651 Polymer engineering laboratory
9841:661 Polymerization reactor engineering
9841:670 Polymer nanocomposites
9841:675 Carbon-polymer nanotechnology
9841:680 Polymer coatings

APPENDIX D

POLYMER ENGINEERING LEADERSHIP

Leadership is what drives any institution. According to Polymer Engineering Department Chair Sadhan Jana, leadership is what has brought the polymer programs at the University of Akron to their level of prominence today. He said, "This leadership has not only attracted the best resources to the university, but also the best minds to the university."[4] James White founded the polymer engineering program at the University of Akron, and while White is still an active faculty member, three others have taken over as positions of chair and institute director. From December 1997 until July 2000, Rudolph J. Scavuzzo, the college's associate dean, was the department's interim chair, and in 2000 Lloyd A. Goettler became its permanent chair. After studying at Cornell University and the University of Delaware, Goettler began a thirty-one-year career at Monsanto Chemical, where he studied injection molding, thermoplastic elastomers, and plastic alloys. During this period, he became friendly with the polymer professors at Akron. After Monsanto, Goettler worked at Solutia in Florida, researching nylon nanocomposites. As a department chair, his goal was to take his industry experience to develop real-world polymer solutions and applications. He said, "I really appreciate the synergy of marrying fundamental study to its practical applications in research work."[5] In 2001, Goettler also became the director of the Institute of Polymer Engineering.

After three years of holding down the positions of both institute director and department chair, Lloyd Goettler decided to relinquish some of his responsibilities. In 2004, Sadhan C. Jana became the new head of the department. Jana had first come to the university in 1998, having earned an undergraduate degree in chemical engineering from the University of Calcutta, a master's degree from the Indian Institute of Technology in Kanpur, and, in 1993, a Ph.D. in chemical engineering from Northwestern University. After completing his studies, he worked on polymer processing properties at GE Corporate Research and Development in New York, where he researched the fundamentals of screw extrusion processes. He came to Akron in part because he wanted to teach, but also because of the potential he saw for its polymer engineering program. Jana believes the polymer programs at the University of Akron have been so successful thanks to leadership that has attracted the best resources and minds in the field to the school. The Polymer Engineering Academic Center has also brought the engineering program into prominence. "The department is very well equipped," Jana says. "I was amazed by the facilities and equipment and plans for the future."[6] Jana wants to be an integral part of this future and the chairmanship has offered him the opportunity to help shape it.

POLYMER ENGINEERING DEPARTMENT

CHAIR	DATES OF SERVICE	CURRENT POSITION
James L. White	7/1/83–12/8/97	Professor of Polymer Engineering; Founding Director of the Institute of Polymer Engineering; Harold A. Morton Professor of Polymer Engineering
Rudolph J. Scavuzzo, Jr.	12/8/97–7/1/00	Interim Department Chair; Professor Emeritus of Polymer Engineering
Lloyd A. Goettler	7/1/00–10/31/04	Retired
Sadhan C. Jana	11/1/04 to Present	Professor of Polymer Engineering

APPENDIX D

DIRECTORS OF THE INSTITUTE OF POLYMER ENGINEERING

NAME	DATES OF SERVICE	CURRENT TITLE
James L White	4/19/1983–12/14/2001	Professor of Polymer Engineering; Founding Director of the Institute of Polymer Engineering; Harold A. Morton Professor of Polymer Engineering
Lloyd A. Goettler	12/14/2001–6/30/2006	Director of the Institute of Polymer Engineering

UNIVERSITY OF AKRON POLYMER ENGINEERING FACULTY, 2006

NAME	TITLE	RESEARCH INTERESTS
Mukerrem Cakmak, Ph.D., Polymer Engineering, University of Tennessee, Knoxville (1984)	Professor of Polymer Engineering, Director (UA Site), Center for Multifunctional Polymer Nanomaterials and Devices (CMPND), Associate Director of Akron Global Polymer Academy (AGPA)	Identification, modeling, and simulation of complex structural mechanisms, particularly stress-induced crystallization that takes place during the course of polymer processing operations of a wide range of polymers subjected to melt as well as rubbery state deformation. The range of materials includes the high temperature thermoplastics and their blends as well as nano particle-filled systems. Of particular interest is the relationship between thermomechanical history applied by fiber spinning, film blowing, biaxial stretching and injection molding and evolved structure and properties. Current activities are focused on real-time measurements of true mechano-optical and mechano-electrical properties of polymers undergoing uni- and biaxial deformation for photonics applications.

NAME	TITLE	RESEARCH INTERESTS
Kevin Cavicchi, PhD, Materials Science and Engineering, University of Minnesota (2003)	Assistant Professor of Polymer Engineering	Thermodynamics of multicomponent polymer systems (block copolymers, graft copolymers, polymer blends); ion-conducting polymers for proton exchange membranes; directed self-assembly of nano-structured and hierarchically ordered materials; anionic polymerization and chemical modification of polymers.
Chang Dae (Paul) Han, Sc. D., Chemical Engineering, Massachusetts Institute of Technology (1964)	Benjamin Franklin Goodrich Professor of Polymer Engineering	Polymer rheology; polymer physics; polymer processing; computational polymer fluid dynamics; miscible and immiscible polymer blends; block copolymers; liquid crystalline polymers; rheology and morphology of structured polymeric liquids.
Hendrik Heinz, Ph. D., Materials Science and Engineering, ETH Zurich (2003)	Assistant Professor	Computer simulation of inorganic-organic interfaces; development of force fields for the quantitative prediction of interfacial properties; biologically inspired and conductive composite materials; structure-property relationships in nanocomposites.
Avraam I. Isayev, Ph. D., Polymer Engineering, Institute of Petrochemical Synthesis of the USSR Academy of Sciences, Moscow (1970)	Distinguished Professor of Polymer Engineering	Polymer processing, rheo-optics, rheology of polymers, oil products, and disperse systems; the injection, co-injection, transfer, compression, and gas-assisted injection molding of polymers; LCP-based self-reinforced or in-situ composites; continuous decrosslinking of thermosets and rubbers and copolymerization of polymer blends with the aid of high power ultrasound; high temperature and high performance nanocomposites; constitutive equations and process modeling.

NAME	TITLE	RESEARCH INTERESTS
Sadhan C. Jana, Ph.D., Chemical Engineering, Northwestern University (1993)	Department Chair; Associate Professor of Polymer Engineering	Chaotic mixing of polymers; conductive polymer compounds; polymer nanocomposite materials; natural fiber composites.
Thein Kyu, Ph.D., Polymer Chemistry, College of Engineering, Kyoto University (1980)	Distinguished Professor of Polymer Engineering	Phase equilibria and kinetics of phase separation in polymer blends; phase transitions in crystalline and liquid crystalline polymers; molecular composites; processing, structure, and properties of ionomer membranes; pattern formation and electro-optical properties of dispersed liquid crystal systems.
Arkadii I. Leonov, Ph.D., Mathematical Physics, Karpov's Physical Chemistry Research Institute, Moscow (1969)	Professor of Polymer Engineering	Thermodynamic modeling, mathematical analysis, and basic experiments for polymer melts, filled polymers, and disperse systems; theoretical/numerical analysis of complex viscoelastic flows; analysis of instabilities and melt fracture phenomena; chemorheology and chemoprocessing of polymers and their filled compositions; theoretical/experimental studies of nematic LC, LCP, LC elastomers, and clay-nanoparticle polymer suspensions; theory of translation/rotation Brownian motion of particles in viscoelastic fluids and dilute polymer solutions; thermodynamic modeling of polymer solids: hybrid models, nonlinear viscoelasticity, necking, and crazing; variational analysis of nonlinear boundary problems for inelastic media.
Kyonsuku Min, Ph.D., Polymer Engineering, University of Tennessee (1984)	Associate Professor of Polymer Engineering	Phase morphology, rheology, and processing of polymer blends; mixing phenonema and process technology for elastomers; reactive blending process.

NAME	TITLE	RESEARCH INTERESTS
Erol Sancaktar, Ph.D., Engineering Mechanics, Virginia Polytechnic Institute and State University (1979)	Professor of Polymer Engineering; Adjunct Professor of Mechanical Engineering	Optimization of adhesive joint geometries, mechanical adhesion; the use of excimer laser in composite material, adhesion, polymer property evaluation, polymer welding, micro-machining applications and nano-lithography; design for new applications with polymer and composites; electrically/electronically conductive polymers/adhesives; ultrasonic welding research.
Mark D. Soucek, Ph.D. Inorganic Chemistry, University of Texas, Austin, (1990)	Associate Professor	Environmentally benign coatings; inorganic/organic hybrid coatings; UV-curable coatings; UV-curable resins; UV-curable powder coatings.
James L. White, Ph.D., Chemical Engineering, University of Delaware (1965)	Professor of Polymer Engineering; Founding Director of the Institute of Polymer Engineering; Harold A. Morton Professor of Polymer Engineering	Twin screw extrusion and batch and continuous mixing; rheology and processing of particle-filled elastomers and plastics; structure development in processing (fiber formation, film extrusion, blow molding, injection molding, etc.); liquid crystalline polymer fluids and polymer blends.
Alan N. Gent, Ph.D., (science), University of London (1949)	Professor Emeritus of Polymer Physics and Polymer Engineering; Research Professor	Deformation and fracture processes in rubbery, crystalline, and glassy polymers.
Nobuyuki Nakajima, Ph.D., Chemistry, Case Institute of Technology (1958)	Professor Emeritus of Polymer Engineering; Research Professor	Characterization of commercial polymers; structure/property relationships of elastomers and polymer melts; behavior of filled polymer systems; polymer blends; polymer processing; powdered rubber; plastisol technology and plastisol foam technology.

NAME	TITLE	RESEARCH INTERESTS
Rudolph J. Scavuzzo, Jr., Ph.D, Mechanical Engineering, University of Pittsburgh (1962)	Professor Emeritus of Polymer Engineering	Dynamics of structures and composites; fatigue and failure of materials; solid mechanics; finite element analysis; seismic and shock response of structures.

APPENDIX E

POLYMER SCIENCE LEADERSHIP

After Maurice Morton stepped down in 1978, eight individuals have been the chair of the Department of Polymer Science. Three of them—Howard Stephens, Irja Piirma, and H. James Harwood—were profiled earlier in this book. The most recent five—Donald McIntyre, Roderic P. Quirk, William J. Brittain, Stephen Z.D. Cheng, and Mark D. Foster—have all served in this leadership position since the formation of the College of Polymer Science and Polymer Engineering, and are profiled below. They have all played an important role in establishing the reputation of the polymer science program today.

The governance structure of the Department of Polymer Sciences includes a rotating chair. Mark Foster, the current chair, has said that this is a critical component in the overall collegiality of the department and enhances the spirit of cooperation among the faculty. According to Foster, this "collaborative nature" and shared leadership structure is one of the key factors responsible for the "health and prominence" of the department. For example, those researchers who are skilled in synthesis can interact with those interested in technology or the physical sciences, and everyone can work together in an interdisciplinary fashion on polymer problems. One central reason is the physical structure of the department itself, its shared space in one building.[7]

After the formation of the polymer college, Donald McIntyre was the first to serve as department chair, a position he held from 1988 until 1994. He had earned his Ph.D. in chemistry in 1954 from Cornell University. Afterwards, he served for two years in the army medical corps and then joined what is now the National Institute of Standards and Technology in 1956 and worked there until he was hired by Maurice Morton in 1966. He said that his philosophy of departmental leadership was to "retain good people, hire additional solid individuals, and then work hard at interfacing with them to establish and maintain a productive, collegial atmosphere."[8]

Roderic P. Quirk succeeded McIntyre and served as chair from 1995 to 1997. Quirk first became interested in science while a high school student in Southfield, Michigan. A graduate of Rensselaer Polytechnic Institute in Troy, New York, he earned his Ph.D., in 1967, in organic chemistry at the University of Illinois. He taught at the University of Arkansas, spending his 1976 sabbatical working with Maurice Morton at the University of Akron. He decided then to shift his academic focus to polymer research and, in 1983, became a professor at the University of Akron. He said that at Akron "you can become all you can be. It's a very strong program, filled with exceptionally talented colleagues."[9] In 1995, he was awarded the university's Outstanding Research Award.[10] Quirk is currently the Kumho Professor and Distinguished Professor of Polymer Science.

After Quirk stepped down, William J. Brittain served as chair from 1998 to 2001. Unlike many of the scientists and engineers in the polymer program who chose their careers at an early age, Brittain initially envisioned a much different life. An aspiring disc jockey, he eventually switched his studies to chemistry and earned his Ph.D. in organic chemistry from the California Institute of Technology in 1982. At Akron, among the goals he envisioned for students who passed through the department while he was chair was a professional transformation. Brittain said, "By the time students have completed their programs, they have been transformed from apprentices to colleagues."[11] He stepped down from the department in 2001 to become Adjunct Professor of Polymer Science and a vice president of Research at Bausch and Lomb headquarters in Rochester, New York.

Stephen Z.D. Cheng took over leadership of the department in 2001 and remained chair until 2005. Cheng had been a physics and math undergraduate at East China Normal University (graduating in 1977) when his father, an electrical engineer, suggested he work with polymers in an applied field. Cheng took his father's advice and earned his master's in polymer engineering at DongHua University in 1981. He then began his doctoral work in polymers at Rensselaer Polytechnic Institute, where he worked on the solid-state physics of macromolecules. After earning his Ph.D. in 1985 in polymer chemistry, he took a position as a research associate in the Advanced Thermal Analysis Laboratory at Rensselaer. When he asked his doctoral advisor about the opening in Akron, he was told, "Apply for the job. Akron is the best place you could possibly go." Cheng arrived on campus in 1987. Soon afterwards, he realized his advisor was correct. Cheng said, "Akron has the most attractive environment I could ever want for my career."[12] He was selected to become the next dean of the College of Polymer Science and Polymer Engineering in 2007.

Mark D. Foster became chair in 2005 and currently holds the position. He earned his Ph.D. in chemical engineering from the University of Minnesota in 1987 and then became a researcher at the Max Planck Institute in Germany for three years, where he studied the structure and dynamics of polymers systems and thin films.[13] In 1990, he was hired by the University of Akron. Foster believes the department's future is bright, saying, "If you would ask me what part of the history of polymer science marks our greatest era, I am inclined to think that we are still heading up. I am excited about what we have been able to do with recent hires in our department and the breadth of research interests that we have." Although he has been at Akron for a relatively short time, compared to some of his colleagues, he has seen a fundamental change in the department, having witnessed a "gradual transformation from a department that just dealt with rubber to a department that now deals with biomolecules, nanotechnology, specialty polymers, and a wide varieties of materials which are polymeric in nature but very different from the materials which were responsible for the original growth of the department."[14]

POLYMER SCIENCE DEPARTMENT

CHAIR	DATES OF SERVICE	CURRENT TITLE
Howard Stephens	1978–1982	Professor Emeritus of Polymer Science
Irja Piirma	1982–8/31/85	Professor Emeritus of Polymer Science
H. James Harwood	9/1/85–8/29/88	Professor Emeritus of Polymer Science and Chemistry
Donald McIntyre	8/30/88–12/31/94	Professor Emeritus of Polymer Science
Roderic P. Quirk	1/1/95–12/31/97	Distinguished Professor and Kumho Professor of Polymer Science

William J. Brittain	1/1/98–8/27/01	Professor of Polymer Science
Stephen Z.D. Cheng	8/28/01–8/31/05	R.C. Musson and Trustees Professor of Polymer Science
Mark D. Foster	9/1/05 to present	Professor of Polymer Science

Like the Department of Polymer Science, the Institute of Polymer Science has also had strong leaders at its helm, though fewer of them, since it is not a rotating position. After the formation of the College of Polymer Science and Polymer Engineering in 1988, Frank Kelley became dean and vacated the position of institute director. Replacing him was Darrell H. Reneker, who began his illustrious polymer career in 1959 after earning his Ph.D. in physics from the University of Chicago. He spent the next ten years at DuPont, where he discovered a previously unknown class of chemical defects that is now referred to as the "Reneker Defect." In 1969, he became a scientist at the polymer division of the National Institute of Standards and Technology, eventually managing a staff of 450. During this period, he was also a member of the Committee on Materials, a White House legislative organization, and played a leading role in developing a new standard for computer databases used in the design and production of polymer composites for aerospace, airplane, and submarine vehicles.[15] He was named institute director at Akron in 1988.

In 1994, Reneker decided he wanted to focus on research and teaching, and resigned as institute director. Frank Harris became the new director of the Institute of Polymer Science in 1994. Harris earned his Ph.D. in chemistry from the University of Iowa in 1968 and has had a distinguished career at the university since his arrival in 1983. He has been the associate director of the National Science Foundation's Center for Molecular and Miscrostructure of Composites and was named the Outstanding Researcher at the university in 1992.[16] After his appointment as institute director, he said that, despite his new responsibilities, he wanted to remain active in his research.[17] Harris stepped down as institute director in 2006 to devote more time to the company he and Stephen Cheng founded, Akron Polymer Systems. Roderic Quirk is currently the interim director of the Maurice Morton Institute of Polymer Science.

APPENDIX E

DIRECTOR OF THE INSTITUTE OF POLYMER SCIENCE

NAME	DATES OF SERVICE	CURRENT TITLE
Maurice Morton	1956–1978	Professor Emeritus of Polymer Science
Frank N. Kelley	1978–1988	Dean of the College of Polymer Science and Polymer Engineering
Darrrell H. Reneker	9/9/1989–9/1/1994	Professor of Polymer Science
Frank Harris	9/1/1994 to 2006	Distinguished Professor of Polymer Science and Biomedical Engineering

UNIVERSITY OF AKRON POLYMER SCIENCE FACULTY, 2006

NAME	TITLE	RESEARCH INTERESTS
Gustavo A. Carri, Ph.D. Polymer Science and Engineering, University of Massachusetts, Amherst (2000)	Associate Professor of Polymer Science	Statistical mechanics of semiflexible polymers; novel mathematical methods in polymer physics; statistical physics of helical polymers; Monte Carlo simulations of helical wormlike polymers; biopolymer dynamics in glassy solvents; molecular dynamics simulations of biopolymers; morphologies of nanocomposites via computer simulations.
Stephen Z.D. Cheng, Ph.D., Polymer Chemistry, Rensselaer Polytechnic Institute (1985)	R.C. Musson and Trustees Professor of Polymer Science	Condensed states in polymeric materials: including transition thermodynamics, kinetics, molecular motion in metastable states, crystal structure and morphology, liquid crystalsand liquid crystalline polymers, surface and interface structures and related phase transitions, phase structures and phase transitions in nano-confined environments, polymer blends and copolymers, high performance polymers, and polymer photonics.

NAME	TITLE	RESEARCH INTERESTS
Scott Collins, Ph.D. Chemistry, University of Calgary	Professor of Polymer Science and Chemistry	Polymerization catalysis using transition metal complexes; synthesis and characterization of organo-transition metal complexes with applications to olefin and acrylate polymerization. Cationic polymerization of isobutene using discrete, organoborane-based Lewis acids; aqueous suspension polymerization of isobutene synthesis of ionomers.
Ali Dhinojwala, Ph.D. Chemical Engineering, Northwestern University (1994)	Professor of Polymer Science	Spectroscopic techniques to study structure and dynamics at surfaces and interfaces; relationship between adhesion, adhesion hysterisis, and friction; structure and dynamics of molecules in confined geometry; surface melting and surface freezing; developing synthetic adhesives using aligned carbon nanotubes inspired by gecko foothairs; electrical and optical applications of aligned carbon nanotubes.
Mark D. Foster, Ph.D., Chemical Engineering, University of Minnesota (1987)	Department Chair; Professor of Polymer Science; Associate Director, Akron Global Polymer Academy	Microstructure and dynamics of polymer systems in thin films and near interfaces, e.g., protein adsorption, surface segregation, and adhesion; x-ray and neutron scattering techniques, especially small-angle scattering and reflectometry for studying these phenomena; scanning nano Raman spectroscopy.
Purushottam D. Gujrati, Ph.D., Physics, Columbia University (1979)	Professor of Physics and Polymer Science	Statistical mechanics and field theory; phase transitions and critical phenomena; renormalization group theory; counting problems; enumerations of graphs; self-avoiding walks; glass transition.

NAME	TITLE	RESEARCH INTERESTS
Gary R. Hamed, Ph.D., Polymer Science, University of Akron (1978)	Professor of Polymer Science	Adhesion; autohesion and tack; brass-rubber adhesion; fracture and durability of rubber; anisotropic rubber; rubber reinforcement.
Frank N. Harris, Ph.D., Chemistry, University of Iowa (1968)	Distinguished Professor of Polymer Science and Biomedical Engineering (part-time)	Condensation polymerization; high performance polymers, including polyimides and polyheterocyclics; structure-property relationships in polymers; polymers for optical, microelectric and fiber applications; controlled-release technology; polymers for medical applications.
Frank N. Kelley, Ph.D., Polymer Chemistry, University of Akron (1961)	Dean, College of Polymer Science and Polymer Engineering; Professor of Polymer Science	Structural characterization and fracture of thermosetting resins; time-dependent fracture of rubbery network polymers; properties of highly filled elastomers.
Joseph P. Kennedy, Ph.D., Biochemistry, University of Vienna (1954)	Distinguished Professor of Polymer Science and Chemistry	Synthesis of well-defined macromolelcules for potential industrial and biomaterial use.
Wayne L. Mattice, Ph.D., Biochemistry, Duke University (1968)	A. Schulman Professor of Polymer Science and Ohio Eminent Scholar	Physical chemistry of macromolecules; dynamics; simulations; photophysics.
George R. Newkome, Ph.D., Organic Chemistry, Kent State University (1966)	Vice President for Research and Dean of the Graduate School; Professor of Polymer Science and Chemistry; Intellectual Property Fellow; James and Vanita Oelschlager Professor of Science and Technology	Synthetic and structural studies in diverse areas of chemistry, supramolecular and nanotechnology; routes to molecular batteries and motors.

NAME	TITLE	RESEARCH INTERESTS
Coleen Pugh, Ph.D, Case Western Reserve University (1990)	Associate Professor of Polymer Science	Synthesis and characterization of novel polymer materials using living polymerizations and new organic reactions: side-chain liquid crystalline polymers, amphiphilic polymers, homopolyrotaxanes and other molecular architectures; fluorinated polymers.
Judit E. Puskas, Ph.D., Plastics and Rubber Technology (Polymer Engineering), Technical University of Budapest (1990)	LANXESS Professor of Polymer Science	Polymers for biomedical application, polymer synthesis, and characterization (SEC expert); polymer structure-property relationships; living carbocationic polymerization; elastomer/rubber science and engineering; macromolecular engineering; polymerization kinetics/mechanisms; women in science and engineering; science and engineering education.
Roderic P. Quirk, Ph.D., Organic Chemistry, University of Illinois (1967)	Kumho Professor and Distinguished Professor of Polymer Science; Interim Director, Maurice Morton Institute of Polymer Science	Anionic polymerization: vinyl, diene, oxirane and methacrylate monomers; functionalization reactions of polymeric carbanions; anionic synthesis of block copolymers and heteroarm star polymers; multifunctional anionic initiators; functionized initiators; lanthanide catalyzed diene polymerization.
Darrell H. Reneker, Ph.D., Physics, University of Chicago (1959)	Professor of Polymer Science	Polymer physics; computer modeling of conformation defects in polymer crystals; morphology and scanning tunneling microscopy of polymers; electrospinning of polymer fibers; electrostriction of polymers; polymer nanofibers.

NAME	TITLE	RESEARCH INTERESTS
Alexei Sokolov, Ph.D., Russian Academy of Sciences (1986)	Thomas A. Knowles Professor of Polymer Science	Nanostructure and dynamics of disordered systems, including polymers and liquids; glass transition; dynamics of bio-polymers; nano-photonics.
Ernst D. von Meerwall, Ph.D., Physics, Northwestern University (1969)	Distinguished Professor of Physics, Chemistry, and Polymer Science; Associate Dean, College of Polymer Science and Polymer Engineering	Pulsed-gradient spin-echo diffusion measurements in polymer and colloidal systems; NMR relaxation studies in polymer solutions and rubbery composites; polymer rheology; fluid dynamics; experiment and computer simulation; thermal, mechanical, and impact measurements in polymer networks; composites.
Shi-Qing Wang, Ph.D., Physics, University of Chicago (1987)	Professor of Polymer Science	Polymer physics and engineering; polymer dynamics and rheology; component dynamics of polymer solutions and blends; flow instabilities and die swell in polymer extrusion and interfacial wall slip; phase separation and diffusion in polymer mixtures; entangle chain dynamics of mixtures—rheology and diffusion behavior.
Marcia E. Weidknecht, B.S., Chemical Engineering, University of New Hampshire (1971)	Instructor of Polymer Science	Solution properties and physical properties of macromolecules; polymerization methods.
William J. Brittain, Ph.D., Organic Chemistry, California Institute of Technology (1982)	Adjunct Professor of Polymer Science; Vice President of Research Bausch and Lomb World Headquarters	Polymer synthesis, nanocomposites; controlled radical polymerization; polymer brushes.

NAME	TITLE	RESEARCH INTERESTS
John E. Frederick, Ph.D., Physical Chemistry, University of Wisconsin (1964)	Professor Emeritus of Polymer Science	Determination of void formation in polymeric systems by light scattering and x-ray scattering; the relationship of viscoelastic properties and scattering. Behavior.
H. James Harwood, Ph. D., Chemistry, Yale University (1956)	Professor Emeritus of Polymer Science	Free radical initiated polymerization; synthetic polypeptides; polymer modification chemistry characterization of polymer microstructure by chemical and spectroscopic methods.
Eberhard A. Meinecke, Dr. Ing., Technische Hochschule, Braunschweig, W. Germany (1950), D. Eng., Braunschweig Institut für Technologie, Germany (1960)	Professor Emeritus of Polymer Science and Mechanical Engineering	Viscoelastic properties of polymers; mechanial design of rubber and plastic components; medical devices.
Donald McIntyre, Ph.D., Chemistry, Cornell University (1954)	Professor Emeritus of Polymer Science	Macromolecular structure, polymeric phase equilibria, and the chemical and physical properties of elastomers, including natural rubbers.
Irja J. Piirma, Ph. D., Polymer Chemistry, University of Akron (1960)	Professor Emeritus of Polymer Science	Emulsion polymerization.
Howard L. Stephens, Ph.D., Chemistry, University of Akron (1960)	Professor Emeritus of Polymer Science and Chemistry	Polymerization, vulcanization, and analysis.

APPENDIX F

MEMBERS OF THE OHIO POLYMER STRATEGY COUNCIL, 2006[18]

Marc Payne	AES
Joe Jacomet	Battelle
Gary Wnek	Case Western Reserve University
Brian Bishop	Demag Plastics Group
Alan Olson	Ferro
Joe Gingo	Goodyear Tire & Rubber
Carol Cartwright	Kent State University
Clark G. Reed	Lockheed Martin Tactical Systems
Bill Rowley	Mercury Plastics
Ron Brown	Milacron
Dorothy Baunach	NorTech
Scott Rickert	NanoFilm
Julian Steinberg	Noveon
Pat Valente—ex-officio	Ohio Department of Development
Tom Brady	Plastic Technologies
Dennis Cocco	PolyOne
Max H. Lewis	Sherwin-Williams
Frank Samuel—ex-officio	State of Ohio, Office of the Governor
Tom Waltermire	OPSC Chairperson & Team NEO
Bob McGrath	The Ohio State University
Luis Proenza	University of Akron
Mickey McCabe	University of Dayton Research Institute

NOTES

INTRODUCTION

1.Brooke Hindle, ed. *America's Wooden Age* (Tarrytown, NY: Sleepy Hollow Restorations, 1975), 3.

2. Herman Mark, "Coming to an Age in Polymers in Science and Technology," *History of Polymer Science and Technology*, ed. Raymond Seymour (New York: Marcel Dekker, 1982), 5.

3. This was based upon a talk given by Clayton F. Ruebensaal, a corporate planning director from Uniroyal. He argued that the new "polymer age" followed three other main periods in the development of rubber: wild rubber, plantation rubber, and synthetic rubber. "New Rubber Age," *Chemical Week* (April 14, 1976).

4. J. Harry DuBois and Frederick W. John, *Plastics*, 6th ed. (New York: Van Nostrand Reinhold, 1981), v.

5. National Research Council, "Growth in Polymer Production," *Akron Beacon Journal*, June 12, 1983.

6. Grecia Matos and Lorie Wagner, "Consumption of Materials in the United States," *Annual Review of Energy and the Environment* 23 (November 1998): 103–122; Kathleen Frankart, "Americans Have a Real Penchant for Polymer," *UA News* (September 1981): 5.

7. David Teegarden, *Polymer Chemistry: Introduction to an Indispensable Science* (Arlington, VA: National Science Teachers Association Press, 2004), xv.

8. Poly Styrene, email message to author, December 11, 2006.

9. She remained somewhat ambivalent about the ultimate good or evil of this Polymer Age. She warned of potential destruction to the natural environment but still called it a "Pandora's Box of hope." Her last line in the song, "I fell from the rayon trees," was prophetic in that she would soon retreat from this persona she created, disband the group she created, X-Ray Spex, and join the Hare Krishna movement.

10. Ed Noga, "A Day in the Life . . ." *Rubber & Plastics News* (March 20, 2006): 12–13.

11. Madeline Goodstein, *Plastics and Polymers Science Fair Projects: Using Hair Gel, Soda Bottles, and Slimy Stuff* (Berkeley Heights, NJ: Enslow Publishers, 2004).

12. Frank Kelley, "The Next Time You String Popcorn, Think of Akron," *Akron Beacon Journal*, (September 25, 1991); Frank Kelley, "What Is a Polymer Anyway?", undated manuscript, College of Polymer Science and Polymer Engineering Archives; Frank Kelley, "Challenges for a Rust-Belt Phoenix," December 16, 1993, Akron Roundtable presentation, College of Polymer Science and Polymer Engineering Archives; "Akron's Future: Polymers," brochure published in 1985, College of Polymer Science and Polymer Engineering Archives.

13. Gary Hamed, interview with the author, March 13, 2008.

14. Steve Love, "Polymers' Pied Pipers," *Akron Beacon Journal*, June 15, 2001.

15. David E. Nye, *American Technological Sublime* (Cambridge, MA: MIT Press, 1994).

16. Meikle, *American Plastic*, 8.

17. Charles E. Carraher, Jr., "History of Polymer Education—USA," in *History of Polymer Science and Technology*, ed. Raymond B. Seymour (New York: Marcel Dekker, 1982), 183.

18. Yasu Furukawa, *Inventing Polymer Science: Staudinger, Carothers, and the Emergence of Macromolecular Chemistry* (Philadelphia: University of Pennsylvania Press, 1997), 2.

19. Frank M. McMillan, *The Chain Straighteners* (London: Macmillan, 1979); Peter J.T. Morris, *Polymer Pioneers: A Popular History of the Science and Technology of Large Molecules* (Philadelphia: Center for History of Chemistry, 1986).

20. John Loadman, *Tears of the Tree: The Story of Rubber—A Modern Marvel* (Oxford: Oxford University Press, 2005); Charles Slack, *Noble Obsession: Charles Goodyear, Thomas Hancock, and the Race to Unlock the Greatest Industrial Secret of the Nineteenth Century* (New York: Hyperion, 2002).

21. Furukawa, *Inventing Polymer Science*, 3.

22. David A. Hounshell and John Kenly Smith, Jr., *Science and Corporate Strategy: DuPont R&D, 1902–1980* (Cambridge: Cambridge University Press, 1988).

23. Mansel G. Blackford and K. Austin Kerr, *BFGoodrich: Tradition and Transformation, 1870–1995* (Columbus: Ohio State University Press, 1996).

24. Steve Love and David Giffels, *Wheels of Fortune* (Akron, OH: University of Akron Press, 1999).

CHAPTER 1

1. John Loadman, *Tears of the Tree: The Story of Rubber—A Modern Marvel* (Oxford: Oxford University Press, 2005), 22.

2. Charles Slack, *Noble Obsession: Charles Goodyear, Thomas Hancock, and the Race to Unlock the Greatest Industrial Secret of the Nineteenth Century* (New York: Hyperion, 2002), 2, 30–31.

3. Charles Goodyear, *The Applications and Uses of Vulcanized Gum-Elastic*, vol. 2 (New Haven, CT: 1855).

4. P.W. Allen, *Natural Rubber and the Synthetics* (New York: John Wiley & Sons, 1972), 31.

5. David E. Nye, *Electrifying America: Social Meanings of a New Technology, 1880–1940* (Cambridge, MA: MIT Press, 1991), 59.

6. James J. Flink, *The Automobile Age* (Cambridge, MA: MIT Press, 1988), 27.

7. William F. Zimmerli, "Rubber Chemistry in Akron University," *India Rubber Review* (March 15, 1915): 151.

8. "Technical Education and the Rubber Industry," *India Rubber Review* (January 1925): 30.

9. Herbert M. Williams, "Editorial," *The Rubber Age* (June 11, 1917): 228.

10. "Hezzleton Erasmus Simmons," *American Chemists and Chemical Engineers*, ed. Wyndham D. Miles (Washington, D.C.: American Chemical Society, 1976), 441.

11. Hezzleton Simmons, "Technical Course in Rubber Manufacture," *The Rubber Age* (July 10, 1917): 307; "University and Factory," *New York Times*, May 9, 1920.

12. "Roguery in Rubber . . . Odd Facts about Akron, Ohio, the 'Rubber City,' " *New York Herald* (July 19, 1904). This article was also reprinted in "Akron a Rubber Center," *India Rubber Review* (August 15, 1904): 250.

13. Mansel G. Blackford and K. Austin Kerr, *BF Goodrich: Tradition and Transformation, 1870–1995* (Columbus: Ohio State University Press, 1996), 14.

14. John Dean Gaffey, *The Productivity of Labor in the Rubber Tire Manufacturing Industry* (New York: Columbia University Press, 1940), 153.

15. "College Man Is Taking Big Part in Tire Making," *Christian Science Monitor*, February 24, 1912.

16. Herbert M. Williams, "Editorial," *The Rubber Age* (June 11, 1917): 228.

17. G. Stafford Whitby, "Thomas Hancock," *Vignettes from the International Rubber Science Hall of Fame*, ed. Barbara Nute Zimmerman (Akron, OH: ACS Rubber Division, 1989): 4–5.

18. Leonard S. Reich, *The Making of American Industrial Research: Science and Business at GE and Bell, 1876–1926* (Cambridge: Cambridge University Press, 1985), 1.

19. Herbert M. Williams, "Editorial," *The Rubber Age* (July 10, 1917): 313.

20. "Dr. Oenslager, 82, Chemist, Is Dead," *New York Times*, February 7, 1956; Blackford and Kerr, *BFGoodrich*, 49.

21. "Diamond Rubber Absorbed," *New York Times*, May 16, 1912.

22. *India Rubber World* (September 1911): 496.

23. Howard I. Cramer, "University Training in Rubber Technology," *India Rubber World* 99, vol. 2 (November 1, 1938): 47.

24. There were some exceptions. The Massachusetts Institute of Technology had offered a thesis option in the chemistry of rubber since the 1890s and by 1938 had supervised approximately seventy senior or graduate theses on the subject of rubber. In London, at nearly the same time that Knight offered his first course in rubber chemistry, Philip Schidrowitz gave six lectures on rubber in the autumn of 1910 at Finsbury Technical College. This college was in operation from 1879–1926 in London, England. Philip Schidrowitz, *Rubber* (London: Methuen, 1911), v.

25. E.V. Osberg, "History of the Division of Rubber Chemistry of the American Chemical Society," *Rubber Chemistry and Technology* (January 1940): 5.

26. Reverend Henry L. Canfield, in George W. Knepper, *Summit's Glory: Sketches of Buchtel College and The University of Akron* (Akron, OH: University of Akron Press, 1990), 9–25.

27. In 1881, Buchtel gave an additional $75,000 to the school, bringing his total donations to nearly $200,000. "College Notes," *New York Times*, June 24, 1881.

28. Major Minor Spicer was Akron's first settler in 1811 and was also a devoted Universalist. Samuel A. Lane, *Fifty Years and Over of Akron and Summit County* (Akron, OH: Beacon Job Department, 1892), 152.

29. "Akron Once on Western Boundary," *Christian Science Monitor*, August 11, 1914.

30. Scott Dix Kenfield, ed., *Akron and Summit County Ohio, 1825–1928* (Chicago: S.J. Clarke Publishing, 1928), 422.

31. William Henry Perrin, ed., *History of Summit County with an Outline Sketch of Ohio* (Chicago: Baskin & Battey Historical Publishers, 1881), 395.

32. George Knepper, *New Lamps for Old* (Akron, OH: University of Akron, 1970), 21.

33. *Annual Catalogue of Buchtel College and Academy*, 1904–05, 31–36, University of Akron Archives.

34. Ralph Waldo Emerson, "The American Scholar," *Essays and English Traits* (New York: P.F. Collier & Son, 1937), 6.

35. Herbert Spencer, *The Study of Sociology* (New York: D. Appleton, 1896), 301–306.

36. Henry M. Leicester, *Historical Background of Chemistry* (New York: Dover Publications, 1956), 218.

37. Christa Jungnickel and Russell McCormmach, *Intellectual Mastery of Nature: Theoretical Physics from Ohm to Einstein*, vol. 2 (Chicago: University of Chicago Press, 1990), 2.

38. "Charles Mellon Knight," *The National Cyclopaedia of American Biography*, 332, 250–251.

39. Spanton, *Fifty Years of Buchtel*, 196.

40. "Slips Back on Memory's Wings to a Denver of Yesterday," *The Daily News: Denver, Colorado*, April 21, 1913.

41. "Charles Mellen Knight," *Who Was Who in America*, vol. 2 (Chicago: A.N. Marquis, 1950), 304.

42. "Buchtel College Notes," *Summit County Beacon*, August 22, 1883.

43. "In Foreign Fields," *Washington Post*, March 5, 1879.

44. Charles Knight to S.C. Cole, 4 February 1920, Box 1/1A/130 #7, University of Akron Archives.

45. A.I. Spanton, ed., *Fifty Years of Buchtel* (Akron, OH: Buchtel College Alumni Association, 1920), 62.

46. Arnold Thackray, Jeffrey L. Sturchio, P. Thomas Carroll, and Robert Bud, *Chemistry in America, 1876–1976* (Dordrecht: D. Reidel, 1985), 29.

47. "Dr. C.M. Knight Has Resigned from Buchtel," 1913, Series E, Box 16, Folder 16, The Knight Family Papers, University of Akron Archives.

48. "Buchtel College Notes," *Summit County Beacon*, November 9, 1881.

49. "Buchtel College in a Bad Way," *New York Times*, March 11, 1896.

50. "Buchtel College Burned," *Washington Post*, December 21, 1899; "Buchtel College Burned," *New York Times*, December 21, 1899.

51. William B. Doyle, *Centennial History of Summit County, Ohio and Representative Citizens* (Chicago: Biographical Publishing Company, 1908), 214.

52. Hezzleton Simmons, "Chemistry at the University of Akron," *The Akron Alumnus* (January 1931): 12.

53. *Annual Catalogue of Buchtel College and Academy*, 1903–04, 42–43, University of Akron Archives.

54. Knepper, *New Lamps for Old*, 94.

55. Charles M. Knight, Diary Entry, 26 November 1910, The Knight Family Papers, Series E: Charles Mellen Knight, Box 5, University of Akron Archives.

56. Charles M. Knight, Diary Entry, 20 May and 23 September 1910, The Knight Family Papers, Series E: Charles Mellen Knight, Box 5, University of Akron Archives.

57. Blackford and K. Austin Kerr, *BFGoodrich: Tradition and Transformation*, 29.

58. Simmons, "The Laboratories and Equipment of the Municipal University of Akron," *Rubber Manufacture: The Cultivation, Chemistry, Testing,*

and Manufacture of Rubber, With Sections on Reclamation of Rubber and the Manufacture of Rubber Substitutes (New York: D. Van Nostrand, 1921), 147.

59. "Chemistry Department Serves a City," *Christian Science Monitor*, May 31, 1923.

60. George W.S. Hayes, "Carnegie Libraries of Ohio," Ohio Preservation Council, at *http://winslo.state.oh.us/services/LPD/presiss_jul03.html.*

61. Charles Olin to H.B. Briggs, 10 April 1905, Box 1/A /130 #7, University of Akron Archives.

62. Charles Olin to Prof. Bates, 1 July 1905, Box 1/A /130 #7, University of Akron Archives.

63. Spanton, *Fifty Years of Buchtel*, 110.

64. Board of Trustees Minutes, 12 May 1909, 1/1A, Location 130, Box 7, University of Akron Archives.

65. "Knight Chemical Laboratory," *The Buchtelite*, April 1909.

66. Spanton, *Fifty Years of Buchtel*, 116.

67. Charles M. Knight, Diary Entry, 12 January 1910, The Knight Family Papers, Series E: Charles Mellen Knight, Box 5, University of Akron Archives.

68. Charles M. Knight, Diary Entry, 18 October 1910, The Knight Family Papers, Series E: Charles Mellen Knight, Box 5, University of Akron Archives.

69. Hezzleton Simmons, "The Laboratories and Equipment of the Municipal University of Akron," *The Rubber Age* (July 10, 1917): 307–308.

70. Maurice Morton, "From Rubber Chemistry to Polymers: A History of Polymer Science at the University of Akron," *Rubber Chemistry and Technology* 62: G25.

71. William C. Geer, *The Reign of Rubber* (New York: Century, 1922), 3.

72. Charles M. Knight, Diary Entry, 5 March 1910, The Knight Family Papers, Series E: Charles Mellen Knight, Box 5, University of Akron Archives.

73. "Knight Chemical Laboratory," *The Buchtelite*, April 1909.

74. Charles M. Knight, Diary Entry, 1 March 1910, The Knight Family Papers, Series E: Charles Mellen Knight, Box 5, University of Akron Archives.

75. "The Chemistry of India Rubber," *Annual Catalogue of Buchtel College and Academy*, 1909–10, 58, University of Akron Archives.

76. Charles M. Knight, Diary Entry, 13 September 1910, The Knight Family Papers, Series E: Charles Mellen Knight, Box 5, University of Akron Archives.

77. "The Chemistry of India Rubber," *Annual Catalogue of Buchtel College and Academy*, 1910–11, 57–58, University of Akron Archives.

78. "Additions to the Library," *The Buchtelite*, September 1910.

79. "President A.B. Church of Buchtel," *New York Times*, 18 November 1912.

80. Board of Directors Minutes, 14 April 1913, University of Akron Archives.

81. A.B. Church to Hezzleton Simmons, 30 July 1906, box 1/1A/130, #7, University of Akron Archives.

82. Charles Bulger, "Hezzleton Simmons," *The Akron Alumnus* (January 1931): 10.

83. Hezzleton Simmons, *Rubber Manufacture*, iii.

84. Hezzleton Simmons, "Chemistry at the University of Akron," *The Akron Alumnus* (January 1931): 12.

85. "Hezzleton Simmons," *The National Cyclopaedia of American Biography* vol. 43, 1961, 559.

86. Hezzleton Simmons, "Chemistry at the University of Akron," *The Akron Alumnus* (January 1931): 16.

87. Herbert M. Williams, "Editorial," *The Rubber Age* (11 June, 1917): 228.

88. Herbert M. Williams, "Editorial," *The Rubber Age* (25 July, 1917): 357.

89. Herbert M. Williams, "Editorial," *The Rubber Age* (10 July, 1917): 313.

90. Advertisement for "A Technical Course in Rubber Manufacture," *The Rubber Age* (25 June, 1917): 268–269.

91. Simmons, *Rubber Manufacture*.

92. Herbert M. Williams, "Editorial," *The Rubber Age* (25 July, 1917): 357.

93. Henry S. Firestone, quoted in "War: Its Effect on the American Rubber Industry," *The Rubber Age* (10 April, 1917): 59.

94. Daniel J. Kevles, *The Physicists: The History of a Scientific Community in Modern America* (Cambridge, MA: Harvard University Press, 1987), 137–138.

95. As part of war preparations, the U.S. government recruited men on college campuses. In 1917, the U.S. government established the Student Army Training Corps (SATC) to prepare students for the military, while also keeping them involved in higher education. On October 1, 1918, over five hundred American colleges held identical ceremonies and inducted 150,000 men. The men previously trained in the Committee on Education's rubber school became the vocational Section B of the SATC, while student members were defined as Section A. *The Official Roster of Ohio Soldiers, Sailors, and Marines in the World War, 1917–18* (Columbus, OH: F.J. Heer, 1926–29).

96. Annual Report 1919, *The Bulletin of the Municipal University of Akron*, 16–17, University of Akron Archives.

97. "Faculty Changes at the University of Akron," *The Alumni Quarterly* (Summer 1919): 1.

98. Spanton, *Fifty Years of Buchtel*, 380.

99. Hezzleton Simmons, "The Laboratories and Equipment of the Municipal University of Akron," *The Rubber Age* (10 July, 1917): 308.

100. "Student Fellowship Open," *The Rubber Age and Tire News* (10 November, 1917): 139.

101. Simmons, "Chemistry at the University of Akron," 16.

102. Annual Reports of the President, Clerk, and Bureau of City Tests, April 1929, University of Akron Archives.

103. "Chemistry Department Serves a City," *Christian Science Monitor*, 31 May, 1923.

104. Barbara N. Zimmerman and Frank N. Kelley, "The Polymer Program at the University of Akron: New Trends, New Directions," *Polymer News* 14 (1989): 311–312.

105. "Round about the Hilltop," *The Akron Alumnus* (December 1929): 7.

106. "Two Foreign Students Drawn to Akron by Chemistry Class," *Akron Beacon Journal*, October 1, 1931.

107. "Japanese Present Swords to Simmons," *Christian Science Monitor*, March 22, 1934.

108. *Akron Beacon Journal*, November 23, 1941.

109. James D. D'Ianni, "Reminiscences on the Centennial Celebration of the Akron U Chemistry Department," June 6, 1985, at *www.chemistry.uakron.edu/PDF/History.pdf*.

110. "Akron U Pioneers in Rubber Study," *Akron Beacon Journal*, February 19, 1939.

111. "J.W. Thomas Dies; Headed Firestone," *New York Times*, November 27, 1951.

112. Alfred Lief, *The Firestone Story: A History of the Firestone Tire & Rubber Company* (New York: McGraw-Hill, 1951), 408.

113. "Firestone Man Retires," *New York Times*, January 23, 1961.

114. "Akron U Pioneers in Rubber Study," *Akron Beacon Journal*, February 19, 1939.

115. Ibid.

116. Howard I. Cramer, "University Training in Rubber Technology," *India Rubber World* 99, vol. 2 (November 1, 1938): 47–49.

117. "Akron U Pioneers in Rubber Study," *Akron Beacon Journal*, February 19, 1939.

118. Dr. Charles M. Knight," Annual Report 1941, 4, University of Akron Archives.

119. "A Worthy Gift," 1937, Series E, Box 16, Folder 16, The Knight Family Papers, University of Akron Archives.

120. "Dr. C.M. Knight Taken by Death," 1941, Series E, Box 16, Folder 16, The Knight Family Papers, University of Akron Archives.

121. Letter from Joseph H. James to A.I. Spanton, Spanton, *Fifty Years at Buchtel*, 199.

CHAPTER 2

1. William Haynes and Ernst A. Hauser, *Rationed Rubber & What to Do about It* (New York: Alfred A. Knopf, 1942), v.

2. American Chemical Society, Plaque on the National Historic Chemical Landmark for the Synthetic Rubber Program, August 1998.

3. "U Research in Rubber Is Vital Task," *Akron Beacon Journal*, October 10, 1952; "University Plays Role in Synthetic Research," *Akron Beacon Journal*, October 22, 1952.

4. The National Historic Chemical Landmark commemorating the United States Synthetic Rubber Program from 1939 to 1945, located outside the Goodyear Polymer Center. This plaque was dedicated by the American Chemical Society on August 29, 1998, and reads, "This site was selected because of its central location and the participation of the rubber research laboratories in the wartime development of synthetic rubber." The plaque also tells the story of the program's significance to the war effort, in part stating: "With U.S. government sponsorship, a consortium of companies involved in rubber research and production united in a unique spirit of technical cooperation and dedication to produce a general-purpose synthetic rubber, GR-S (Government Rubber-Styrene), on a commercial scale. In Akron and other U.S. locations these companies, in collaboration with a network of researchers in numerous government, academic, and industrial laboratories, developed and manufactured in record time enough synthetic rubber to meet the needs of the U.S. and its allies during World War II: The Firestone Tire & Rubber Company, The B.F. Goodrich Company, The Goodyear Tire & Rubber Company, Standard Oil Company of New Jersey, and United States Rubber Company."

5. "Rubber Act of 1948," U.S. Senate Report No. 1015, March 15, 1948.

6. "Jones and Meyer Put on Sideshow of Fisticuffs," *Los Angeles Times* , April 11, 1942.

7. "Jesse Jones Shakes Eugene Meyer; Eye-Glasses Broken in Encounter," *New York Times* (10 April 1942): 1; William M. Tuttle, "The Birth of an Industry: The Synthetic Rubber 'Mess" in World War II," *Technology and Culture* (1981): 35–67.

8. "Fifth of Rubber Industry's Production Is Defense Work," *Wall Street Journal*, May 16, 1941.

9. Otto Scott, Bill Mulligan, Joseph Del Gatto, and Ken Allison, "The Division of Rubber Chemistry: Catalyst of an Industry Part III," *Rubber World* (December 1966): 90.

10. Henry E. Rose, "Key to Large-Scale Synthetic Rubber Output Is Materials for New Plants," *Wall Street Journal*, June 4, 1942.

11. Peter Morris, *The American Synthetic Rubber Research Program* (Philadelphia: University of Pennsylvania Press, 1989), 1.

12. Frank A. Howard, *Buna Rubber: The Birth of an Industry* (New York: D. Van Nostrand, 1947): 215.

13. Vernon Herbert and Attillo Bisio, *Synthetic Rubber: A Project That Had to Succeed* (Westport, CT: Greenwood Press, 1985); Maurice Morton, "History of Synthetic Rubber," *History of Polymer Science and Technology*, ed. Raymond B. Seymour (New York: Marcel Decker, 1982), 225–238.

14. R.H. Markham, "Rubber Making, American Way: Big Machines Controlled by Little People," *Christian Science Monitor*, June 21, 1945.

15. Maurice Morton, "G. Stafford Whitby," *American Chemists and Chemical Engineers*, ed. Wyndham D. Miles, (Washington, D.C.: American Chemical Society, 1976), 505.

16. President Harry S. Truman, "Recommendations Concerning the Synthetic Rubber Industry in the United States," U.S. House of Representatives Document No. 326, January 14, 1952.

17. Letter from Norman P. Auburn to General E.L. Ford, in Board of Directors Minutes, 18 June 1953, University of Akron Archives.

18. Morris, *The American Synthetic Rubber Research Program*, 7.

19. Edwin C. Heinke, "Plan Research Rubber Bureau for University," newspaper article, in Box 4/8A/8, University of Akron Archives.

20. Letter from A.L. Viles (general manager of the Rubber Manufacturers of America) to George Zook (president of the University of Akron), 22 May 1929, Box 4/8A/8, University of Akron Archives.

21. "Memorandum to the Board of the Rubber Association of America Proposing the Establishment of an Institute of Rubber Technology at the University of Akron," undated, Box 4/8A/8, University of Akron Archives.

21. R.P. Dinsmore, "Institute for Rubber Research Proposal," undated, Box 4/8A/8, University of Akron Archives.

22. Mansel G. Blackford and K. Austin Kerr, *BFGoodrich: Tradition and Transformation, 1870–1995* (Columbus: Ohio State University Press, 1996), 111.

23. Howard Wolf and Ralph Wolf, *Rubber: A Story of Glory and Greed* (New York: Covici, Friede Publishers, 1936), 460.

24. Letter from Frank Seiberling to George Zook, 8 January 1930, Box 4/8A/8, University of Akron Archives.

25. Morris, *The American Synthetic Rubber Research Program*, 8.

26. "The University's War Effort," 1942 Annual Report, 11, University of Akron Archives.

27. Morton, "G. Stafford Whitby," 503–505.

28. Wolf, *Rubber*, 379.

29. Whitby, quoted in Otto Scott, "A Man for All Seasons," *Rubber World* (December 1965): 20.

30. For a review of some of Whitby's publications from these years, see: G. S. Whitby, *India Rubber Journal* 48 (1914): 805; B.J. Eaton and Whitby, *The Agricultural Bulletin of the Federated Malay States* 5 (1917): 124; Whitby, *The Agricultural Bulletin of the Federated Malay States* 6 (1918): 374, 513.

31. G. Stafford Whitby, *Pilgrim Soul: Poems and Stories by Wynne Whitby* (London: Linden Press, 1961), 7.

32. Ibid., 38.

33. Ibid., 43.

34. G. Stafford Whitby, *Plantation Rubber and the Testing of Rubber* (London: Longmans, Green, 1920).

35. There he was an outspoken advocate for the importance of chemistry to Canadian industry as a whole. For example, in a *Wall Street Journal* article, Whitby discussed the significance of a new process of manufacturing ethylene from natural gas. "Process Provides Use for Waste Gas," *Wall Street Journal*, June 3, 1931.

36. G. Stafford Whitby, "Some Fundamental Rubber Problems," *Rubber Chemistry and Technology* 4 (1931): 465.

37. Maurice Morton, "George Stafford Whitby," *Rubber Chemistry and Technology* (December 1972): G78-G85.

38. J.A. Pickard to G. Stafford Whitby, 8 November 1938, Whitby Collection, Box 1, University of Akron Archives.

39. "Chemical Society Choice for Rubber Group Head," *New York Times*, December 1948.

40. Maurice Morton, "From Rubber Chemistry to Polymers: A History of Polymer Science at the University of Akron," *Rubber Chemistry and Technology* 62: G28.

41. Whitby to William Cullen, 19 September 1942, Whitby Collection, Box 1, University of Akron Archives.

42. Whitby, *Pilgrim Soul*, 114.

43. Whitby and Simmons, "Vulcanization Accelerators," *Industrial and Engineering Chemistry* 17 (1925): 931–933.

44. Wolf, *Rubber*, 345.

45. G. Stafford Whitby, "Reflections on Rubber Research," *Industrial and Engineering Chemistry* 47, vol. 5 (1955): 806.

46. "Students at U 'Double Up' to Conserve Rubber," *Akron Beacon Journal*, January 9, 1942.

47. Harold J. Harwood, "Memories of G. Stafford Whitby," email message to the author, February 13, 2006.

48. Ibid.

49. James D. D'Ianni, "Reminiscences on the Centennial Celebration of the Akron U Chemistry Department," June 6, 1985, at *www.chemistry.uakron.edu/pdf/history.pdf*.

50. Whitby's correspondence with Dinsmore on this "Information Exchange with Canada," detailed specific technical efforts. In March 1943, Whitby sent Dinsmore reports by the National Research Council in Ottawa on a new method for preparing butadiene and an apparatus for the rapid analysis of

butadiene samples. Whitby to Dinsmore, 11 March 1943. The Whitby Collection, Box 1, University of Akron Archives. R.P. Dinsmore to Whitby, 30 December 1942. The Whitby Collection, Box 1, University of Akron Archives.

51. "Doctor G. Stafford Whitby," 25 September 1942, Minutes of the Board of Directors, University of Akron Archives.

52. "Akron U Conducts Extensive Synthetic Rubber Research," *Akron Beacon Journal*, October 31, 1943.

53. Board of Directors Minutes, 12 April 1954, University of Akron Archives.

54. Herbert and Bisio, *Synthetic Rubber*, 151.

55. Whitby, N. Wellman, V.W. Floutz, and H.L. Stephens, "Polamine-Activated Polymerization," *Industrial and Engineering Chemistry* 42, vol. 3 (1950): 451–452.

56. Maurice Morton, "From Rubber Chemistry to Polymers: A History of Polymer Science at the University of Akron," *Rubber Chemistry and Technology* 62: G28.

57. Herbert and Bisio, *Synthetic Rubber*, 152.

58. Maurice Morton, "G. Stafford Whitby," *Vignettes from the International Rubber Science Hall of Fame*, ed. Barbara Nute Zimmerman (Akron, OH: ACS Rubber Division, 1989), 111.

59. Whitby, editor in chief, *Synthetic Rubber* (New York: J. Wiley, 1954).

60. Morton, "G. Stafford Whitby," *Vignettes*, 112.

61. "Honor Rubber Scientist," *New York Times*, April 25, 1954.

62. Whitby, "Melting Pot," *Christian Science Monitor*, December 27, 1966.

63. "Biographical Sketch of Retired Faculty Member to Be Included in Board Minutes—G. Stafford Whitby," 18 September 1954, Board of Director's Minutes, University of Akron Archives.

64. "Akron U to Aid Research Plan," *Akron Beacon Journal*, December 26, 1943.

65. Herbert and Bisio, *Synthetic Rubber*, 149.

66. Morris, *The American Synthetic Rubber Research Program*, 117.

67. Ibid., 70–83.

68. "College Heads Ask Anti-War Act Now," *New York Times*, May 18, 1934.

69. Howard, *Buna Rubber*, 223.

70. "President Simmons to Spend More Time in Washington," 25 September 1942, Minutes of the Board of Directors, University of Akron Archives.

71. "Secret and Confidential, War Production Board," Minutes of Meeting #1 in Akron, 28–29 December 1942, College of Polymer Science and Polymer Engineering Archives.

72. "Rubber Institute," 13 February 1942, Minutes of the Board of Directors, University of Akron Archives.

73. "Rubber Research," 1943 Annual Report, 8, University of Akron Archives; Rubber Reserve Company—Synthetic Rubber," 26 February 1943, Minutes of the Board of Directors, University of Akron Archives.

74. "Promotion of an Institute for Rubber Research," 25 October 1943, Minutes of the Board of Directors, University of Akron Archives.

75. "Rubber Reserve Company," 10 December 1943, Minutes of the Board of Directors, University of Akron Archives.

76. "Agreement with the Defense Plant Corporation," 14 January 1944, Minutes of the Board of Directors, University of Akron Archives.

77. "Government Laboratories," 1944 Annual Report, University of Akron Archives.

78. "Rubber Reserve Co. Contract—Pilot Plant & Evaluation Laboratory," 11 February 1944, Minutes of the Board of Directors, University of Akron Archives.

79. "Akron U Conducts Rubber Research," *Akron Beacon Journal*, October 18, 1955.

80. "Dewey to See Dedication of Synthetic Rubber Plant," *Akron Beacon Journal*, June 15, 1944.

81. "Offers to Sell GSA Rubber Lab," *Wall Street Journal*, April 22, 1957.

82. "Suggestions for Continuation of Research in Synthetic Rubber," 18 June 1953, Minutes of the Board of Directors, University of Akron Archives.

83. "Appointments for Government Laboratories," 14 January 1944, Minutes of the Board of Directors, University of Akron Archives.

84. "Preliminary Service Rates," 18 August 1950, Minutes of the Board of Directors, University of Akron Archives.

85. Whitby to B.G. Labbe, 6 December 1945, The Whitby Collection, Box 1, University of Akron Archives.

86. Whitby to B.G. Labbe, 11 December 1945, The Whitby Collection, Box 1, University of Akron Archives.

87. Herbert and Bisio, *Synthetic Rubber*, 150.

88. Hezzleton Simmons, "Government Laboratory," 1946 Annual Report, University of Akron Archives.

89. Ibid.

90. "Rubber Act of 1948," U.S. Senate Report No. 1015, March 15, 1948.

91. Hezzleton Simmons, "The Government Laboratories," 1947 Annual Report, University of Akron Archives.

92. "U Continues Rubber Study," *Akron Beacon Journal*, September 24, 1949.

93. "Approval of 1949–50 Operating Budget," 23 September 1949, Minutes of the Board of Directors, University of Akron Archives.

94. President Harry S. Truman, "Recommendations Concerning the Synthetic Rubber Industry in the United States," U.S. House of Representatives Document No. 326, January 14, 1952.

95. Hezzleton Simmons, "The Government Laboratories," 1950 Annual Report, 11, University of Akron Archives.

96. James W. Schade, "The Government Laboratories," 1953 Annual Report, 25–26, University of Akron Archives.

97. W.K. Taft, "The Government Laboratories," 1952 Annual Report, 31, University of Akron Archives.

98. W.K. Taft, "The Government Laboratories," 1952 Annual Report, 31–32, University of Akron Archives.

99. Schade to Auburn, 20 June 1952, Minutes of the Board of Directors, University of Akron Archives.

100. "Government Laboratories," 18 January 1954, Minutes of the Board of Directors, University of Akron Archives.

101. "Government Laboratories," 20 June 1952, Minutes of the Board of Directors, University of Akron Archives.

102. President Dwight D. Eisenhower, "Recommendations Relative to the Rubber Act of 1948, as Amended, relating to the Synthetic-Rubber Program," U.S. House of Representatives Document No. 123, April 14, 1953.

103. Milt Freudenheim, "Continue U Rubber Studies," *Akron Beacon Journal*, May 14, 1953.

104. Norman P. Auburn, "Synthetic Rubber Research," 28 October 1953, Minutes of the Board of Directors, University of Akron Archives.

105. "Synthetic Lab Sale to U Favored," *Akron Beacon Journal*, June 19, 1953.

106. W.K. Taft, "The Government Laboratories," 1955 Annual Report, 41, University of Akron Archives.

107. Hope MacLeod, "U Seeks U.S. Labs," *Akron Beacon Journal*, March 24, 1954.

108. "Firms Back Rubber Lab Sale to U, Urge U.S. to Take Step," *Akron Beacon Journal*, January 3, 1955.

109. Letter, 22 December 1954, Minutes of the Board of Directors, University of Akron Archives.

110. "U.S. Keeps Rubber Lab at Akron U," *Akron Beacon Journal*, January 25, 1955.

111. This committee included: Arthur C. Cope, Edwin R. Gilliland, and William A. Krebs of MIT; Farrington Daniels of the University of Wisconsin; William H. Davis, the former economic stabilizer; Dean Joseph C. Elgin of Princeton University; Paul D. Foote of Gulf Research and Development Company; President David D. Henry of the University of Illinois; Frank A. Oward from Standard Oil Development; Chancellor Lawrence A. Kimpton and Warren C. Johnson of the University of Chicago. "Study Plan in Rubber Research," *Akron Beacon Journal*, September 29, 1955. Matt Freudenheim, "Urges U.S. to End AU Lab Contracts," *Akron Beacon Journal*, December 19, 1955.

112. "Special Group Studies Action on Rubber Lab," *Akron Beacon Journal*, January 7, 1956.

113. "AU to Drop Government Rubber Project," *The Buchtelite*, February 28, 1956.

114. Terry Turner, "Heed Advice of 5-Man Committee," *Akron Beacon Journal*, February 16, 1956.

115. "Special Group Studies Action on Rubber Lab," *Akron Beacon Journal*, January 7, 1956.

116. "Synthetic Lab's Fate Months Off," *Akron Beacon Journal*, November 1, 1956.

117. "Editorials," *Rubber Age* (May 1956): 293.

118. "Washington at Work," *Wall Street Journal*, July 20, 1956.

119. "Wilbeth Rd. Plant Sale Authorized," *Akron Beacon Journal*, July 27, 1956.

120. "Offers to Sell GSA Rubber Lab," *Wall Street Journal*, April 22, 1957.

121. "Dr. Simmons' Death Shocks Many Friends," *The Akron Alumnus* (Spring 1955): 4.

122. Hezzleton Simmons, "Chemistry at the University of Akron," *The Akron Alumnus* (January 1931): 16.

CHAPTER 3

1. Peter Hayes, *Industry and Ideology: IG Farben in the Nazi Era* (Cambridge: Cambridge University Press, 1987), xi.

2. James White, interview with the author, October 24, 2006. White was a student of Mark's.

3. Victor F. McElhenry, "At 80, Dr. Mark Is Honored for Pioneering in Polymers," *New York Times*, May 3, 1975.

4. Mansel G. Blackford and K. Austin Kerr, *BFGoodrich: Tradition and Transformation, 1870–1995* (Columbus: Ohio State University Press, 1996), 231.

5. Maurice Morton, "Institute of Rubber Research," Annual Reports of the President and the Treasurer for the Year 1956, University of Akron Archives.

6. Norman P. Auburn, "Russia's 'New Look' in Higher Education," *The Akron Alumnus* (Winter 1956): 3; Maurice Morton, "A Shot in the Arm for U.S. Science," *Akron Beacon Journal*, November 17, 1957.

7. Norman P. Auburn, "Sputniks Demand Careful Analysis; Are Educators Demanding Quality?" *The Akron Alumnus* (Winter 1957–1958): 2.

8. Norman P. Auburn, "Education Under Scrutiny," report to the board of directors, *The University of Akron Annual Report* (1957–1958), 3.

9. Norman P. Auburn, "Development of 'E-Bomb' at Alumni Conference," *The Akron Alumnus* (Summer 1958): 2.

10. Norman P. Auburn, "Research Institute, Building Addition Are New Advances," *The Akron Alumnus* (Spring 1956): 4.

11. Yasu Furukawa, "Polymer Chemistry," *Science in the Twentieth Century*, ed. John Krige and Dominique Pestre (Australia: Harwood Academic Publishers, 1997), 547.

12. Yasu Furukawa, *Inventing Polymer Science: Staudinger, Carothers, and the Emergence of Macromolecular Chemistry* (Philadelphia: University of Pennsylvania Press, 1997), 1–6.

13. Ibid., 12, 66.

14. A. Fredga, "Les Prix Nobel en 1953," *Pioneers in Polymer Science*, ed. Raymond B. Seymour (Dordrecht: Kluwer Academic Publishers, 1989), 22.

15. "Dr. Hermann Staudinger Dies; Chemist Won 1953 Nobel Prize," *New York Times*, September 10, 1965; William Quarles, "Herman Staudinger: Thirty Years of Macromolecules," *Journal of Chemical Education* 28 (1951): 120–122.

16. Furukawa, *Inventing Polymer Science*, 59.

17. Ibid., 7.

18. Staudinger, *From Organic Chemistry to Macromolecules: A Scientific Autobiography on My Original Papers* (New York: Wiley-Interscience Publishers, 1970), 77.

19. Furukawa, *Inventing Polymer Science*, 12, 66.

20. G. Stafford Whitby, "Herman Staudinger (1881–1965)," *Rubber Chemistry and Technology* 40 (June 1967): xvi-xxiii.

21. Victor F. McElhenry, "At 80, Dr. Mark is Honored for Pioneering in Polymers," *New York Times*, May 3, 1975.

22. Staudinger, *From Organic Chemistry to Macromolecules*, 6.

23. Victor F. McElhenry, "At 80, Dr. Mark is Honored for Pioneering in Polymers," *New York Times*, May 3, 1975.

24. Peter Hayes, *Industry and Ideology: IG Farben in the Nazi Era*, xi.

25. Furukawa, *Inventing Polymer Science*, 77, 79.

26. G. Allan Stahl, "Development of a Modern Polymer Theory," *Pioneers in Polymer Science*, ed. Raymond B. Seymour (Dordrecht: Kluwer Academic Publishers, 1989), 24.

27. Eiji Ochiai, "Profile of Dr. Staudinger," *Kobunshi* 3 (1954): 223–226, translated by Furukawa, *Inventing Polymer Science*, 76.

28. C.S. Marvel, "Wallace Hume Carothers—Innovator, Motivator, Pioneer," *Pioneers in Polymer Science*, 127–143.

29. Wallace Carothers, "Studies on Polymerization and Ring Formation: An Introduction to the General Theory of Condensation Polymers," *Journal of the American Chemical Society* 51 (August 1929): 2548.

30. David A. Hounshell and John Kenly Smith, Jr., *Science and Corporate Strategy: DuPont R&D, 1902–1980* (Cambridge: Cambridge University Press,

1988), 221; Alfred D. Chandler, Jr. and Stephen Salsbury, *Pierre S. Du Pont and the Making of the Modern Corporation* (New York: Harper & Row, Publishers, 1971).

31. Furukawa, *Inventing Polymer Science*, 95.

32. Staudinger, in Furukawa, *Inventing Polymer Science*, 142.

33. Maurice Morton, interview, December 16, 1992, College of Polymer Science and Polymer Engineering Archives.

34. Herman Mark, "Coming to an Age in Polymers in Science and Technology," *History of Polymer Science and Technology*, ed. Raymond B. Seymour (New York: Marcel Dekker, 1982), 5.

35. "Dr. W. H. Carothers, Chemist, Found Dead, Co-Inventor of a Synthetic Rubber Believed a Suicide by Poison in Philadelphia," *The New York Times*, April 30, 1937.

36. Furukawa, *Inventing Polymer Science*, 198.

37. Hounshell and Smith, *Science and Corporate Strategy: DuPont R&D, 1902–1980*, 296.

38. "Dr. Mark Is Honored by Noted Scientists," *New York Times*, May 28, 1965.

39. "Nobel Award Due Expert on Plastics," *Washington Post*, November 4, 1953.

40. "German, Dutchman Given Nobel Prizes," *Washington Post*, November 5, 1953.

41. Herman F. Mark, "Hermann Staudinger—Father of Modern Polymer Science," *Pioneers in Polymer Science*, 93–109.

42. "Chemical Award Won by Rubber Research," *New York Times*, May 13, 1947.

43. "Paul Flory, Nobel Laureate in Chemistry," *Chicago Tribune*, September 12, 1985.

44. Paul Flory, *Principles of Polymer Chemistry* (Ithaca, NY: Cornell University Press, 1953).

45. "Paul Flory, 75, Nobel Laureate in Chemistry, Rights Activist, Dies," *Washington Post*, September 12, 1985.

46. Charles E. Carraher, Jr., "History of Polymer Education—USA," *History of Polymer Science and Technology*, 173.

47. "Educational Courses in Rubber Chemistry and Technology," *Rubber Red Book* (1955–56): 981–989.

48. David Ewers, "Rubber Education: A Short History, A Long Future," *Rubber World* (October 1989): 114.

49. Robert Pierson, "Maurice Morton, Charles Goodyear Medalist—1985," *Rubber Chemistry and Technology, Rubber Reviews for 1985* (July-August 1985): G73.

50. Don McClure, *Akron Beacon Journal*, October 30, 1955.

51. Maurice Morton, interview, December 16, 1992, College of Polymer Science and Polymer Engineering Archives.

52. David J. Ewers and Frank N. Kelley, "Maurice Morton, 1913–1994," *Rubber Chemistry and Technology* (May-June 1995): G28.

53. Ron Kirksey, "UA Polymer Program Founder Morton, 81," *Akron Beacon Journal*, March 24, 1994.

54. Maurice Morton, "Rubber Enters the Polymer Age," *Rubber Chemistry and Technology, Rubber Reviews for* 1985 (July-August 1985): G75.

55. Ibid.

56. Stuart Drown, "City Shaped By Giants in Polymers," *Akron Beacon Journal*, July 13, 1994.

57. "Polymer Lecture Group to Meet Here June 9," *Akron Beacon Journal*, June 1, 1947.

58. "Polymer Lecture Group Announces Program," *Akron Beacon Journal*, August 17, 1947.

59. Allied Akron Architects prepared the initial plans and specifications, and Clemmer Construction was awarded the general construction bid. "Awarding of Contracts for New Chemistry Building," 20 May 1949, Minutes of the Board of Directors, University of Akron Archives.

60. "Advertisement Bids of Authorized New Chemistry Building," 23 March 1949, Minutes of the Board of Directors, University of Akron Archives.

61. The keynote speaker at the ceremony was Harry L. Fisher, a B.F. Goodrich research chemist; his address was entitled "Rubber Plus Chemistry." "Dedication of the New Chemistry Building," 21 September 1950, Minutes of the Board of Directors, University of Akron Archives.

62. "The Knight Chemical Laboratory," Dedication Booklet, 18 January 1951, College of Polymer Science and Polymer Engineering Archives.

63. "Rubber Research Laboratory," 20 June 1952, Minutes of the Board of Directors, University of Akron Archives.

64. "Department of Rubber Research," 23 January 1953, Minutes of the Board of Directors, University of Akron Archives.

65. Maurice Morton, interview, December 16, 1992, College of Polymer Science and Polymer Engineering Archives.

66. "New President Is Named by University of Akron," *New York Times*, June 30, 1951.

67. George Knepper, *New Lamps for Old* (Akron, OH: University of Akron, 1970), 364.

68. Ibid., 293–298.

69. "Scientific Research Program Urged at Akron U," *Akron Beacon Journal*, February 17, 1952.

70. Text of Eisenhower's opening campaign reelection speech, Frances Payne Bolton Papers, Box 51, Folder 869, Western Reserve Historical Society. For historical context of the Cold War, see Walter LaFeber, *America, Russia, and the Cold War, 1945–1996*, 8th ed. (New York: McGraw-Hill, 1997).

71. Jesse Shera, "A Plan for the Creation of a National Center for the Coordination of Scientific and Technical Information," February 1958, Series RG-90/F-49, University of Pittsburgh Archives.

72. G. Mennen Williams, Democratic governor of Michigan, in James T. Patterson, *Grand Expectations: The United States, 1945–1974* (Oxford: Oxford University Press, 1996), 418.

73. For the response to Sputnik from University of Akron students and faculty, see Miriam Wein, "Sputnik—Has It Changed Your Brand of Breakfast Cereal Yet?" *The Buchtelite*, October 25, 1957.

74. Loren R. Graham, *Science in Russia and the Soviet Union: A Short History* (New York: Cambridge University Press, 1993); Zhores A. Medvedev, *Soviet Science* (New York: W.W. Norton, 1978). See also: Sonya G. Machelson, "Soviet Scientific and Technical Propaganda," *American Documentation* 11 (1960): 237–240; A.L. Mackay, "Sources of Russian Scientific Information," *ASLIB Proceedings* 6, vol. 2 (May 1954): 101–110; A.E. Stubbs, "The Dissemination of the Knowledge of Soviet Scientific Work in Western Countries," *ASLIB Proceedings* 9, vol. 11 (November 1957): 333–340.

75. G. Kharatyan, *Vestnik Statistiki* (1962), as translated in "Scientific Personnel: Number, Composition, and Training," *Soviet Review* 3, vol. 9 (September 1962): 38.

76. G. Pascal Zachary, *Endless Frontier: Vannever Bush, Engineer of the American Century* (New York: The Free Press, 1997), 1.

77. Stuart W. Leslie, *The Cold War and American Science: The Military-Industrial-Academic Complex at MIT and Stanford* (New York: Columbia University Press, 1993), 1–6.

78. "Auburns Leave for Europe," *The Buchtelite*, July 5, 1956.

79. Norman P. Auburn, "Russia's 'New Look' in Higher Education," *The Akron Alumnus* (Winter 1956): 3–12.

80. Norman P. Auburn, "Education Under Scrutiny," report to the board of directors, *The University of Akron Annual Report* (1957–1958), 4.

81. Norman P. Auburn, "Research Institute, Building Addition Are New Advances," *The Akron Alumnus* (Spring 1956): 4.

82. Maurice Morton, "The University of Akron, The Institute of Rubber Research," 16 March 1955, Minutes of the Board of Directors, "Office of Rubber Research," University of Akron Archives.

83. Norman P. Auburn, "A Second Look at Russian Higher Education," *The Akron Alumnus* (Special Issue 1959): 2–15.

84. "Auburn Itinerary," *The Buchtelite*, April 8, 1959.

85. "Dr. Auburn: 'Russians Progressing,'" *The Buchtelite*, May 29, 1959.

86. Norman P. Auburn, "Development of 'E-Bomb' at Alumni Conference," *The Akron Alumnus* (Summer 1958): 2.

87. Auburn, in *New Lamps for Old*, 300.

88. Maurice Morton, "A Shot in the Arm for U.S. Science," *Akron Beacon Journal*, November 17, 1957.

89. *The Philadelphia Enquirer*, August 17, 1984.

90. "Paul Flory, Nobel Laureate in Chemistry," *Chicago Tribune*, September 12, 1985.

91. Stuart Drown, "City Shaped by Giants in Polymers," *Akron Beacon Journal*, July 13, 1994.

92. Maurice Morton, "Recommendations for the Establishment of a Rubber Research Institute at the University of Akron," February 1956, College of Polymer Science and Polymer Engineering Archives.

93. Maurice Morton, "A Shot in the Arm for U.S. Science," *Akron Beacon Journal*, November 17, 1957.

94. Maurice Morton, "Rubber Research Laboratory," Annual Report to the President, 1955, University of Akron Archives.

95. "Office of Rubber Research," 5 January 1955, Minutes of the Board of Directors, University of Akron Archives.

96. In 1954, Fred J. Kovac and Kermit C. Breach were awarded the *Rubber Age* prize for their master's thesis on "A Compounding Study of Butyl Rubber," 27 May 1955, Minutes of the Board of Directors, "Office of Rubber Research," University of Akron Archives.

97. Maurice Morton, interview, December 16, 1992, College of Polymer Science and Polymer Engineering Archives.

98. Maurice Morton, "Rubber Research Laboratory," Annual Report to the President, 1955, University of Akron Archives.

99. Maurice Morton, "Rubber Research Institute," 25 January 1956, Minutes of the Board of Directors, "Office of Rubber Research," University of Akron Archives.

100. "Chemistry Department, *Annual Report of the University of Akron* (1952–1953), 33, University of Akron Archives; "Dr. Sumner Named 'Outstanding Man' by Akron JayCees," *Akron Alumnus* (Spring 1958): 6.

101. Norman P. Auburn, "Education Under Scrutiny," report to the Board of Directors, *The University of Akron Annual Report* (1957–1958), 5.

102. "Chemistry Lab Work Taught by Films," *Akron Alumnus* (Spring 1959): 3.

103. "Indicators of pH," produced by the University of Akron for Coronet Instructional Films, 1959, University of Akron Archives.

104. Craig S. Thompson, "College Subs Films for Chemistry Labs, Finds They May Raise Grades, Save Money," *Wall Street Journal*, March 19, 1959.

105. Maurice Morton to Norman Auburn, 7 February 1956, College of Polymer Science and Polymer Engineering Archives.

106. Maurice Morton, "Recommendations for the Establishment of a Rubber Research Institute at the University of Akron," February 1956, College of Polymer Science and Polymer Engineering Archives.

107. Ibid.

108. Norman Auburn to Maurice Morton, 8 March 1956, College of Polymer Science and Polymer Engineering Archives.

109. Maurice Morton, "The University of Akron, The Institute of Rubber Research," 16 March 1955, Minutes of the Board of Directors, "Office of Rubber Research," University of Akron Archives.

110. "AU Forms Rubber Institute," *Akron Beacon Journal*, March 23, 1956.

111. Maurice Morton, "The University of Akron, The Institute of Rubber Research," 16 March 1955, Minutes of the Board of Directors, "Office of Rubber Research," University of Akron Archives.

112. "Akron University Establishes Institute for Rubber Research," *Rubber Age* (April 1956): 115; Maurice Morton, "Institute of Rubber Research," Annual Reports of the President and the Treasurer for the Year 1956, University of Akron Archives.

113. Morton to Auburn, 28 March 1956, College of Polymer Science and Polymer Engineering Archives; Maurice Morton, "Institute of Rubber Research," Annual Reports of the President and the Treasurer for the Year 1956, University of Akron Archives.

114. Maurice Morton, letter sent throughout the rubber industry and academic institutions, 17 May 1956, College of Polymer Science and Polymer Engineering Archives.

115. Institute of Rubber Research, promotional brochure, 1956, College of Polymer Science and Polymer Engineering Archives.

116. Thomas Sumner, "Chemistry Department," *The University of Akron Annual Report* (1958–1959), 13–14.

117. "Microscope Given AU by Goodyear," *Akron Beacon Journal*, October 23, 1957.

118. "Now We Can See A Molecule In Akron U's Research Program Labs," *The Akron Alumnus* (Winter 1957–58): 6.

119. "Akron University Reports on New Rubber Research Institute," *Rubber Age* (May 1956): 298.

120. Maurice Morton, "Institute of Rubber Research," *The University of Akron Annual Report* (1958–1959), 41.

121. "Akron University Awarded Contract," *Rubber Age* (September 1958): 1035.

122. Maurice Morton, interview, December 16, 1992, College of Polymer Science and Polymer Engineering Archives.

123. Ibid.

124. Steven Rattner, "In Akron's Retread Ritual, Pickets Need No Signs," *New York Times*, May 31, 1976.

125. "Ph.D. in Polymer Chemistry Marks University Milestone," *The Akron Alumnus* (Summer 1956): 6.

126. Maurice Morton, "Recommendations for the Establishment of a Rubber Research Institute at the University of Akron," February 1956, College of Polymer Science and Polymer Engineering Archives.

127. Norman P. Auburn, Annual Reports of the President and the Treasurer for the Year 1952, University of Akron Archives.

128. Auburn, quoted in Steve Love and David Giffels, "Akron's New Source of Pride: Men of Science, Vision Have Made UA a Symbol," *Akron Beacon Journal*, December 14, 1997.

129. Ron Kirksey, "UA Polymer Program Founder Morton, 81," *Akron Beacon Journal*, March 24, 1994.

130. Maurice Morton, interview, December 16, 1992, College of Polymer Science and Polymer Engineering Archives.

131. Norman P. Auburn, "Doctor of Philosophy Program in Polymer Chemistry," 19 May 1956, Minutes of the Board of Directors, University of Akron Archives.

132. Norman P. Auburn, remarks at the "Twenty-Fifth Anniversary Celebrating the Institute of Polymer Science," 17 September 1981, College of Polymer Science and Polymer Engineering Archives.

133. Auburn, "Doctor of Philosophy Program in Polymer Chemistry," 19 May 1956, Minutes of the Board of Directors, University of Akron Archives.

134. "Offers Polymer Chemistry Degree," *Rubber Age* (August 1956): 834.

135. "Ph.D.—Akron," *Akron Beacon Journal*, May 31,1956.

136. June 1956, Minutes of the Board of Directors, University of Akron Archives.

137. "Chemistry Department," *The University of Akron Annual Report* (1955–1956), 32, University of Akron Archives.

138. June 1956, Minutes of the Board of Directors, University of Akron Archives.

139. Maurice Morton, "Institute of Rubber Research," Annual Reports of the President and the Treasurer for the Year 1959, University of Akron Archives.

140. "First Doctorates Conferred by the University," *The Akron Alumnus* (July 1959).

141. "AU Accredited to Grant Doctorate in Chemistry," *The Akron Alumnus* (Spring 1960): 2.

142. Maurice Morton, "Ralph Milkovich—A Memoir," Presented at the Symposium Banquet, nineteenth Akron Polymer Conference, 21 June 1988, College of Polymer Science and Polymer Engineering Archives.

143. M. Szwarc, "Living Polymers. Their Discovery, Characterization, and Properties," *Journal of Polymer Science: Part A: Polymer Chemistry* 36 (1998): xiv.

144. Ralph Milkovich, "Monodisperse Difunctional Polymers" (Ph.D. thesis, University of Akron, 1959).

145. Moshe Levy, "'Living Polymers'—50 Years of Evolution," at *http://www.weizmann.ac.il/ICS/booklet/18/pdf/levy.pdf.*

146. "The Ralph Milkovich Memorial Symposium, Nineteenth Annual Polymer Conference, Advances in Heterophase Polymers (Akron, Ohio: 1988).

147. "Bostick Named Winner of Annual RUBBER AGE Award," *Rubber Age* (August 1959): 810.

148. "Edgar Earl Bostick," *Akron* 17, vol. 2 (Spring 2003).

149. "Chemistry Department, *University of Akron Annual Report* (1955–1956), 29, University of Akron Archives.

150. "University of Akron Awards First Ph.D. Degrees," *Rubber Age* (August 1959): 810.

151. Norman P. Auburn, "Education Under Scrutiny," report to the board of directors, *The University of Akron Annual Report* (1957–1958), 3–4.

152. Herman Mark to Maurice Morton, 17 December 1964, College of Polymer Science and Polymer Engineering Archives.

153. Maurice Morton to Herman Mark, 21 December 1964, College of Polymer Science and Polymer Engineering Archives.

CHAPTER 4

1. Other new specialties included: data processing, cryogenics, and high energy physics. "New Specialties," *Time*, October 19, 1962.

2. Maurice Morton to Norman Auburn, 19 October 1962, College of Polymer Science and Polymer Engineering Archives.

3. "JFK Is Assassinated by Sniper in Dallas," *Akron Beacon Journal*, November 22, 1963.

4. Norman P. Auburn, handwritten opening remarks, 23 November 1963, College of Polymer Science and Polymer Engineering Archives.

5. Norman P. Auburn, remarks at the "Twenty-fifth Anniversary Celebration of the Institute of Polymer Science," College of Polymer Science and Polymer Engineering Archives.

6. Maurice Morton to Norman P. Auburn, 14 May 1958, College of Polymer Science and Polymer Engineering Archives.

7. Maurice Morton to Norman P. Auburn, 14 May 1958, College of Polymer Science and Polymer Engineering Archives. Maurice Morton to the Fiftieth Anniversary Committee, 26 June 1958, College of Polymer Science and Polymer Engineering Archives.

8. George W. Ball, "Minutes of the Committee for the Fiftieth Anniversary Observance of the Teaching of Rubber Chemistry at the University of Akron," 17 September 1958, College of Polymer Science and Polymer Engineering Archives.

9. Barbara Zimmerman, ed. *Vignettes from the International Rubber Science Hall of Fame (1958–1988): 36 Major Contributors to Rubber Science, A Biographical Collection* (Akron, OH: The Rubber Division of the American Chemical Society, 1989), 1.

10. Norman Auburn to Maurice Morton, 11 April 1958, College of Polymer Science and Polymer Engineering Archives.

11. Maurice Morton, undated, typed notes that he called "Behind the Scenes," describing the events of the fiftieth anniversary, College of Polymer Science and Polymer Engineering Archives.

12. Ibid.

13. Maurice Morton to Paul Flory, 26 June 1958, College of Polymer Science and Polymer Engineering Archives.

14. Alan Waterman, quoted in "Constant Research Key to Future of US Science," *Akron Beacon Journal*, October 4, 1958.

15. Norman P. Auburn to Maurice Morton, 6 October 1958, College of Polymer Science and Polymer Engineering Archives.

16. Maurice Morton address to the Board of Directors, 19 December 1962, College of Polymer Science and Polymer Engineering Archives.

17. Ibid.

18. H. James Harwoord, correspondence with the author, February 2, 2006.

19. "Reaching for the 'Highest Highs' in Life," *Akron Polymer* (Spring 1991): 12.

20. "Piirma Named Professor Emeritus," *Akron Polymer* (Fall 1990): 4.

21. "Dr. Alan Gent Faculty Profile," *Akron Polymer* (Fall 1992): 8.

22. Alan Gent, interview with the author, June 21, 2006.

23. "Rubber Springs for Cars?" *Akron Beacon Journal*, February 28, 1965.

24. "Dr. Eberhard A. Meinecke Faculty Profile," *Akron Polymer* (Fall 1992): 9.

25. This team included Paul C. Cross, a physical chemist and president of the Mellon Institute; Malcolm Dole, a physical chemist and chairman of Northwestern University's materials research center; Martin A. Elliott, a gas

engineer and vice president of Illinois Institute of Technology; John F. Meck, an attorney and treasurer and vice president of Dartmouth College; Lyle W. Phillips, a physicist and director of the NSF division for undergraduate education in science; and Richard A. Carrigan, a research chemist from the NSF. "AU Facing 'Test' on Polymer Science," *Akron Beacon Journal*, November 16, 1965.

26. "Dr. Don McIntyre Focuses on 'Unity, Liberty, and Charity,'" *Akron Polymer* (Spring 1993): 11.

27. Lewis J. Fetters, "2000 a Banner Year for University Alumnus," *Polymer* (Spring 2001): 4.

28. "Dr. Joseph P. Kennedy's Brush with Communists Propels Career Forward," *Akron Polymer* (Spring 1992): 11.

29. "AU Adds Polymer Science Doctorate," *Akron Beacon Journal*, June 5, 1964.

30. The rubber industry leaders included C.F. Gibbs and J.E. Jansen from Goodrich; J. D'Ianni and A.J. Gracia from Goodyear; Glen Alliger and B.L. Johnson from Firestone; G.H. Swart and E.E. Gruber from General Tire; and W.E. Bissinger and N.V. Seeger from PPG.

31. Morton to Auburn, 27 May 1964, College of Polymer Science and Polymer Engineering Archives.

32. Morton, "It's 'Polymer Science' Now at Akron U," *Akron Beacon Journal* (24 October 1965).

33. "An Interdisciplinary Program Leading to the Doctor of Philosophy Degree in Polymer Science," 6 April 1964, College of Polymer Science and Polymer Engineering Archives.

34. Morton to Paul H. Johnson, 20 November 1964, College of Polymer Science and Polymer Engineering Archives.

35. Morton, "Present Status and Future Potential of the Institute of Polymer Science," 6 December 1966, 3, College of Polymer Science and Polymer Engineering Archives.

36. Ibid., 5.

37. George W. Knepper, *New Lamps for Old* (Akron: University of Akron, 1970), 345–349.

38. Morton, "An Ohio Resource—Proposal for the Expansion of the Institute of Rubber Research," 28 October 1962, College of Polymer Science and Polymer Engineering Archives.

39. Knepper, *New Lamps for Old* , 345–349.

40. Ibid., 349.

41. Auburn to dedication committee, 9 November 1966, College of Polymer Science and Polymer Engineering Archives.

42. Norman Auburn to Maurice Morton, "Dedication, Science and Engineering Center," 11 November 1966, College of Polymer Science and Polymer

Engineering Archives. Other members of the planning committee included Senior Vice President and Provost D.J. Guzzetta, Professor George W. Knepper, Coleman Major, M.J. Rzasa, Samuel Selby, and George W. Ball.

43. Maurice Morton to D.J. Guzzetta, 8 March 1967, College of Polymer Science and Polymer Engineering Archives.

44. Minutes, Dedication Committee for the Auburn Science and Engineering Center, 17 March 1968, College of Polymer Science and Polymer Engineering Archives.

45. Maurice Morton to D.J. Guzzetta, 8 March 1967, College of Polymer Science and Polymer Engineering Archives.

46. The celebration took place at the University Club.

47. A.B. Japs to Morton, 4 August 1967, College of Polymer Science and Polymer Engineering Archives; Morton to Waldo Semon, 16 June 1967, College of Polymer Science and Polymer Engineering Archives.

48. Minutes, Dedication Committee for the Auburn Science and Engineering Center, 9 January 1968, College of Polymer Science and Polymer Engineering Archives.

49. Whitby to Morton, 26 July 1967, handwritten letter, College of Polymer Science and Polymer Engineering Archives.

50. "They Recall Critical Hour . . . And the Birth of a Growing Industry," *Akron Beacon Journal*, April 20, 1968.

51. "Governor to Help Dedicate Auburn Center," *Akron Beacon Journal*, April 10, 1968.

52. These included Arthur Tobolsky of Princeton University, who spoke on the molecular theory of polymers; James M. McKelvey, the dean of the college of engineering at Washington University, who spoke on polymer engineering; and Phillip K. Reily of the U.S. Department of Commerce, who gave a presentation on the state technical services program. "The Norman Paul Auburn Science and Engineering Center Dedicatory Symposium," 19 April 1968, College of Polymer Science and Polymer Engineering Archives.

53. George W. Ball, "Dedication Events, Auburn Center," undated interoffice memo, College of Polymer Science and Polymer Engineering.

54. Auburn to Morton, 22 April 1968, College of Polymer Science and Polymer Engineering Archives.

55. W.O. Baker to Morton, 7 May 1968, College of Polymer Science and Polymer Engineering Archives.

56. E.V. Osberg, "History of the Division of Rubber Chemistry of the American Chemical Society," *Rubber Chemistry and Technology* (January 1940): 1–2.

57. Ibid., 5.

58. *India Rubber World* (December 1909).

59. Ben Kastein, "Historical Highlights of the Rubber Division ACS," *Rubber Division* 75th *Anniversary* (1984): 43.

60. Otto Scott, Bill Mulligan, Joseph Del Gatto, and Ken Allison, "The Division of Rubber Chemistry: Catalyst of an Industry Part II," *Rubber World* (November 1966): 86.

61. Bruno Latour and Steve Woolgar, *Laboratory Life: The Construction of Scientific Facts* (Princeton, NJ: Princeton University Press, 1986), 45–53.

62. Josephine A. Cushman, "A Special Library for the Rubber Industry," 1920, University of Akron Archives.

63. "The Rubber Division Library," *Rubber Division* 75th *Anniversary* (1984): 59.

64. Ralph E. Wolf, "The Library of the Division of Rubber Chemistry, A.C.S.," *Rubber World* (February 1952): 582–584.

65. "Summary of Development of Rubber Division Library," in President's Report of November 25, 1953, University of Akron Archives.

66. "Report of Library Committee to Directors of Division of Rubber Chemistry," in President's Report of November 25, 1953, University of Akron Archives.

67. Clinebell was kept busy responding to 130 reference questions, making 557 interlibrary loans, and compiling fifteen bibliographies, including a thirty-three-page annotated bibliography on the *Engineering Development of Rubber, 1951–1952*. "Rubber Division Library," in Annual Reports of the President 1952, University of Akron Archives.

68. "AU Names Acting Chief for Information Service," *Akron Beacon Journal*, December 15, 1965.

69. Morton to Ian MacGregor, 11 October 1962, College of Polymer Science and Polymer Engineering Archives.

70. Otto Scott, et. al., "The Division of Rubber Chemistry: Part IV," 70.

71. Charles Rader, interview with the author, June 14, 2006.

72. Marge Bauer, in "The Rubber Division Office," *Rubber Division* 75th *Anniversary* (1984): 58.

73. Otto Scott, Bill Mulligan, Joseph Del Gatto, and Ken Allison, "The Division of Rubber Chemistry: Catalyst of an Industry Part IV," *Rubber World* (January 1967): 70.

74. Lundberg to Morton, 6 June 1961, College of Polymer Science and Polymer Engineering Archives.

75. The weeklong series of courses, with over twenty different speakers, provided an important forum for the interchange of creative ideas in the field. Warren W. Leigh, the dean of the College of Business Administration, supervised this program, with Morton serving as educational consultant. "33 Dealers Sign for AU Tire Course," *Akron Beacon Journal*, January 22, 1957.

76. This was a one-week intensive course that cost $150 and was open to advanced scientists and engineers in the field. There was space for twenty-five participants, and they were exposed to twenty-eight hours of lectures and fifteen hours of laboratory work. The lectures were given by professors at the university as well as experts from the rubber industry. "The Chemistry and Physics of Elastomers," brochure, August 1959, University of Akron Publications, 1959–1960, University of Akron Archives.

77. "Akron University Completes Second Elastomers Course," *Rubber Age* (July 1957): 651; "AU Rubber Course Has Wide Appeal," *Akron Beacon Journal*, August 26, 1959.

78. The program coordinators were a diverse interdisciplinary group including: Norman Washburn (sociology); Ray Sandefur (speech); Paul Weidner (political science); Cathryn Taliaferro (English); Edgar Toppin (history); James Dunlap (speech); William Mavrides (television coordinator); Robert Kirkhart (clinical psychologist); and Ida Goshkin from the Akron Public Library. "'A Chance to Learn' Starts at AU Monday for Unionists," *Akron Beacon Journal*, February 2, 1961.

79. "A Chance to Learn," educational brochure, University of Akron Publications, 1959–1960, University of Akron Archives.

80. "200 Sign Up for 'Chance to Learn,'" *Akron Beacon Journal*, February 13, 1961.

81. Over 270 rubber employees attended the lectures, with a series fee of $15. The first lecturer was William Denton of Goodrich, who presented an introduction to tire development. "Start AU Lectures on Tire Making," *Akron Beacon Journal*, February 20, 1961; "Tire Class at AU Hears McCormick," *Akron Beacon Journal*, March 4, 1961.

82. These lectures were held in the John S. Knight Auditorium, which could accommodate an audience of two hundred people. The first lecture was given by Glen Alliger of Firestone on the topic of "An Introduction to Elastomer Vulcanization." The course was advertised by stating that a background in rubber compounding or chemistry was not necessary. "Rubber Unit, AU Present Lectures," *Akron Beacon Journal*, February 1, 1962.

83. "Polymer Lecturers Announced," *Akron Beacon Journal*, September 18, 1963.

84. Morton to Lundberg, 22 June 1961, College of Polymer Science and Polymer Engineering Archives.

85. Preface to the Correspondence Course Study Guide, August 1962, College of Polymer Science and Polymer Engineering Archives.

86. Maurice Morton, ed. *Introduction to Rubber Technology* (New York: Reinhold Publishing, 1959), iii.

87. The fifteen weeks were broken down into the following topics: history and summary of rubber technology; natural rubber; vulcanization and accel-

erators of vulcanization; antioxidants and antiozonants; rubber plasticizers, softeners and extenders; carbon black; nonblack compounding ingredients; butadiene-styrene rubbers; nitrile and polyacrylate; butyl rubber; neoprene and hypalon synthetic rubber; polysulfide polymers and silicone rubbers; reclaimed rubber; latex, sponge, and foam; physical testing; and processing equipment. Lundberg to Officers of the Rubber Groups, 20 April 1962, College of Polymer Science and Polymer Engineering Archives. Rubber Division Education Committee to Rubber Group Members, 20 July 1962, College of Polymer Science and Polymer Engineering Archives.

88. "Registration Statistics for Introduction to Rubber Technology Correspondence Course," 16 October 1962, College of Polymer Science and Polymer Engineering Archives.

89. The Rubber Division recognizes industry achievements through nine major awards and two undergraduate scholarships, including the Charles Goodyear Medal, the Melvin Mooney Distinguished Technology Award, the Fernley H. Banbury Award, the Sparks-Thomas Award , the Chemistry of Thermoplastic Elastomers Award, the George Stafford Whitby Award for Distinguished Teaching and Research, the Arnold Smith Special Service Award, the Distinguished Service Award, and the Certificate of Special Appreciation. "Our Vision," Rubber Division of the American Chemical Society, at *http://www.rubber.org/vision_mission_antitrust.htm.*

90. Charles Rader, interview with the author, June 14, 2006.

91. Charles Rader, interview with the author, June 14, 2006.

92. "Norman Auburn's Day," *Akron Beacon Journal*, May 6, 1962.

93. "Auburn, Key '50s, '60s UA President Dies," *Akron Beacon Journal*, July 22, 2003.

94. "UA Pioneer," *Akron Update* (13 April 1994): 1.

95. "UA Honors an Early Leader in the Polymer Sciences," *Akron Beacon Journal*, February 7, 1993.

96. Reneker, quoted in Ron Kirksey, "UA Honors an Early Leader in the Polymer Sciences," *Akron Beacon Journal*, February 7, 1993.

97. Mark Auburn quoted in Ron Kirksey, "Morton Institute," *Akron Beacon Journal*, December 3, 1992.

98. Don McIntyre to J. Alex Morton, 1 April 1994, College of Polymer Science and Polymer Engineering Archives.

99. R.J. Chang to Frank Kelley, 21 April 1994, College of Polymer Science and Polymer Engineering Archives.

100. "Death Claims Polymer Pioneer Dr. Maurice Morton," University of Akron, undated news release, College of Polymer Science and Polymer Engineering Archives.

101. "Maurice Morton, 80, Polymer Expert," *Cleveland Plain Dealer*, March 24, 1994.

CHAPTER 5

1. *The University of Akron Annual Report*, (1984–1985), 7, University of Akron Archives.

2. Dale Allen, "Polymer Science Building: A Symbol of Vision and Resilience," *Akron Beacon Journal*, March 17, 1991.

3. Stephen Franklin, "As Big Wheels Exit, Akron Eyes New Treads," *Chicago Tribune*, July 28, 1991.

4. Joyce Dyer, *Gum-Dipped: A Daughter Remembers Rubber Town* (Akron, OH: University of Akron Press, 2003), 178–180.

5. Steve Love and David Giffels, *Wheels of Fortune* (Akron, OH: University of Akron Press, 1999), 177, 178, 190–193.

6. Ibid., 194.

7. "Akron Plans to Join High-Tech Club," *Miami Herald*, May 9, 1983.

8. Cristal Williams Walker, "The Ups and Downs of Downtown," *Akron Beacon Journal*, December 11, 1989.

9. Jeffrey L. Rodengen, *The Legend of Goodyear: The First 100 Years* (Florida: Write Stuff Syndicate, 1997), 180.

10. Paul Dickson and William D. Hickman, *Firestone: A Legend. A Century. A Celebration* (New York: Forbes Custom Publishing, 2000), 108.

11. Frank Kelley, interview with the author, September 15, 2006.

12. Love and Giffels, "Akron's New Source of Pride, Men of Science, Vision, Have Made UA a Symbol," *Akron Beacon Journal*, December 14, 1997.

13. "They're Probing Rubber's Secrets," *Akron Beacon Journal*, March 20, 1960.

14. Frank Kelley, interview with the author, September 15, 2006.

15. Robert A. Dubick and Isadore Newman, "Interview with Frank N. Kelley," *Ohio Journal of Science* 96 (1996): 72–75.

16. Frank Kelley, interview with the author, September 15, 2006.

17. Alan Gent, interview with the author, June 20, 2006.

18. D.J. Guzzetta, "Initial Action for the 1980s," *University of Akron Annual Report*, 1980–81.

19. Barbara Schechter, "Akron U Institute Is Lauded, Polymer Science Celebrates Twenty-Fifth Anniversary," *Akron Beacon Journal*, September 18, 1981.

20. Twenty-fifth Anniversary Celebration booklet, 17 September 1981, College of Polymer Science and Polymer Engineering Archives.

21. "Polymer Science Ph.D. Program Reaches 25-yr. Mark," *Modern Plastics* (November 1981): 116.

22. Herman Mark to Frank Kelley, 21 September 1981, College of Polymer Science and Polymer Engineering Archives.

23. Paul Flory to Maurice Morton, 8 April 1981, College of Polymer Science and Polymer Engineering Archives.

24. Frank Kelley to Carl Snyder of Goodyear, 17 July 1981, College of Polymer Science and Polymer Engineering Archives. Similar letters were sent to S.L. Aggarwal (General Tire), A.G. Altenau (Firestone), and Robert Fawcett (B.F. Goodrich).

25. "A Plan for the '80s," The Institute of Polymer Science at the University of Akron, *Annual Report to the President*, 1979–80, University of Akron Archives.

26. William Muse, interview with the author, September 7, 2006.

27. William Muse, quoted in Maureen Brown, "Muse Will Stress Quality at Akron U," *Akron Beacon Journal*, September 6, 1985.

28. *The University of Akron Annual Report*, (1984–1985), 13, University of Akron Archives.

29. William Muse, interview with the author, September 7, 2006.

30. Ibid.

31. Martin Kenney, ed. *Understanding Silicon Valley: The Anatomy of an Entrepreneurial Region* (Stanford, CA: Stanford University Press, 2000); Susan Rosegrant and David R. Lampe, *Route 128: Lessons from Boston's High-Tech Community* (New York: Basic Books, 1992).

32. "Akron Jockeys for Position in High-Tech Job Scramble," *Akron Beacon Journal*, January 16, 1984.

33. "Summit 21—A Concept for New Growth," 1981, College of Polymer Science and Polymer Engineering Archives.

34. Barbara Schechter, "Polymer Tech Center Is Envisioned in Area," *Akron Beacon Journal*, December 2, 1981.

35. Editorial, "Polymer Science Center Idea Should Be Pursued," *Akron Beacon Journal*, December 9, 1981.

36. Donald Iannone, quoted in "Is Akron-Cleveland Coalition Ahead," *Akron Beacon Journal*, June 12. 1983.

37. Frank Kelley, "Challenges for a Rust-Belt Phoenix," 16 December 1993, Akron Roundtable presentation, College of Polymer Science and Polymer Engineering Archives.

38. Robert Mercer, quoted in "The University of Akron Research Plaza," College of Polymer Science and Polymer Engineering Archives.

39. "Polymer Task Force Resource Book," 1984, College of Polymer Science and Polymer Engineering Archives.

40. Steve Love and David Giffels, "The Polymer Generation Outgrowth of Rubber Offers New Opportunities," *Akron Beacon Journal*, December 21, 1997.

41. Bill Sloat, "Nobel Winner Cites Akron's Potential," *Akron Beacon Journal*, October 6, 1984.

42. Eric Baer, quoted in Webb Shaw, "Ohio Is Investing in the Future with Research, Training Centers," *Akron Beacon Journal*, May 20, 1985.

43. Marilyn Geewax, "Will Polymer Research Shape Akron's Future?" *Akron Beacon Journal*, February 12, 1983.

44. Joe Coates, quoted in "High-Technology Comeback Is Akron's Economic Goal," *Akron Beacon Journal*, June 12, 1983.

45. Jack L. Dustin, James L. Shanahan, and Peter E. Challis, "Polymer Technology, Innovation, and Economic Development," 1, College of Polymer Science and Polymer Engineering Archives.

46. Maureen Brown, "Polymer Study Optimistic, But Further Detail Urged," *Akron Beacon Journal*, December 6, 1985.

47. Jack L. Dustin, James L. Shanahan, and Peter E. Challis, "Polymer Technology, Innovation and Economic Development: Linking the Futures of the Industry and Northeast Ohio" (Washington, D.C.: R. Leo Penne Associates, 1985), iii.

48. Tom Sawyer, "Akron's Future: Polymers," brochure published in 1985, College of Polymer Science and Polymer Engineering Archives.

49. William McKnight, quoted in "Research Bright Spot for Area," *Akron Beacon Journal*, January 6, 1985.

50. "A Proposal for an Advanced Technology Application Center in Polymers," April 1984, College of Polymer Science and Polymer Engineering Archives.

51. "A Proposal for an Advanced Technology Application Center in Polymers," April 1984, College of Polymer Science and Polymer Engineering Archives.

52. Sharon Coleman, "Case President Urges Polymer Patience," *Akron Beacon Journal*, September 21, 1984.

53. "Polymer Partnership Official," *Akron Beacon Journal*, July 12, 1984; "$4.1 Million Polymer Study Grants OK'd," *Akron Beacon Journal*, August 1, 1984.

54. Along with EPIC, eight other Edison Technology Centers were established in Ohio in the 1980s. These included the Edison Welding Institute (EWI), which was located in Columbus and specialized in welding or joining technology to decrease manufacturing costs and improve industrial products. The Edison Materials Technology Center (EMTEC) was located in Dayton and focused on finding new ways to develop stronger and lighter materials, bind them together better, and improve materials processing. The Edison Industrial Systems Center (EISC), based in Toledo, focused on large-scale systems integration such as automated vision systems. The Edison Biotechnology Center (EBTC) in Cleveland coordinated its resources to develop biomedical products for commercial application. The Edison Animal Biotechnology Center (EABC) in Athens was another biotechnology group devoted to applications in gene transfer, animal biology, and molecular and

cellular biology. The Institute of Advanced Manufacturing Sciences (IAMS) in Cincinnati explored ways to improve the fundamental aspects of manufacturing. Finally, the Cleveland Advanced Manufacturing Program (CAMP) offered a comprehensive manufacturing support program. The state's leaders hoped that the combined efforts of all these centers, scattered throughout Ohio, would significantly elevate the scientific and technical climate in Ohio. Bob Paynter and Mary Grace Poidomani, "Polymer Project Funded," *Akron Beacon Journal*, July 11, 1984. "Ohio's Thomas Edison Program," brochure, College of Polymer Science and Polymer Engineering Archives.

55. *The University of Akron Annual Report*, (1984–1985), 7, University of Akron Archives.

56. Frank Kelley, interview with the author, September 21, 2006.

57. "The Polymer Push," *The University of Akron Annual Report*, (1988–1989), *Akron* (Fall 1989).

58. Anne Hiltner, quoted in Webb Shaw, "From Old Warehouse to 3-Floor Miniplant," *Akron Beacon Journal*, May 20, 1985.

59. Frank Kelley quoted in "A High Calling," *Akron Beacon Journal*, August 27, 1985.

60. Dwight Woodward, "Edison Project Pools Expertise to Aid Economy," *Akron Beacon Journal*, December 13, 1987.

61. Mery deHaas Tomsick, "New Idea Struggles to Find Funds to Make It Work," *Crain's Cleveland Business*, October 21, 1991.

62. "Producers Share Research to Whip the Competition," *New York Times*, March 5, 1989.

63. Robert Fernandez, "Polymer Technology Promoter Referees Two Universities' Rivalry, He's Keeping the Lines Open," *Akron Beacon Journal*, June 18, 1990.

64. Maureen Brown, "Polymer Study Optimistic, But Further Detail Urged," *Akron Beacon Journal*, December 6, 1985.

65. Norman P. Auburn, remarks at the "Twenty-fifth Anniversary Celebration of the Institute of Polymer Science," College of Polymer Science and Polymer Engineering Archives.

66. William Muse, interview with the author, September 7, 2006.

67. H. James Harwood, interview with the author, October 7, 2005.

68. Frank Kelley quoted in Marcia Myers, "Akron U Wants Polymer Programs Combined," *Akron Beacon Journal*, December 1, 1987.

69. Stephen Z.D. Cheng, interview with the author, June 21, 2006.

70. Joseph P. Kennedy, interview with the author, June 20, 2006.

71. Frank Kelley, interview with the author, September 21, 2006.

72. Frank Kelley quoted in "Polymer Expert Joins AU Program," *Akron Beacon Journal*, November 15, 1985.

73. William Hershey, "Akron U, KSU Win State Grants," *Akron Beacon Journal*, May 11, 1984.

74. "The Polymer Push," *The University of Akron Annual Report*, (1988–1989), *Akron* (Fall 1989).

75. Wayne Mattice, quoted in "Polymer Expert Joins AU Program," *Akron Beacon Journal*, November 15, 1985.

76. "Eby to Be Second Eminent Scholar," *Akron Polymer* 1, vol. 1 (Spring 1990): 1.

77. Mark Foster, interview with the author, October 7, 2005.

78. Cristal Williams Walker, "UA Professor Wins NASA's Snoopy Award, Recognizes Work on Correcting Faulty Boosters," *Akron Beacon Journal*, March 30, 1991.

79. Robert Fernandez, "Korean Tire Maker Endows AU College," *Akron Beacon Journal*, May 2, 1989.

80. Maureen Brown, "Akron U Polymer Unit to Join Space Research Project," *Akron Beacon Journal*, August 23, 1985.

81. Susan Smith, "Did O-Ring Study Cost AU a Grant," *Akron Beacon Journal*, September 29, 1988.

82. Barbara N. Zimmerman, "New College, Modern Polymer Research Facility at the University of Akron Strengthen Polymer Program," letter from Barbara Zimmerman to James H. Pflanz, III, 27 February 1990, College of Polymer Science and Polymer Engineering Archives.

83. The previous expansion into the renovated Whitby Hall laboratories in 1975 had helped accommodate some new growth. Built with a $3.2 million grant from the state of Ohio, Whitby Hall increased the Institute of Polymer Science from 40,000 square feet to 65,000 square feet. Maurice Morton said the addition had "undoubtedly given us the largest and best equipped academic facility in polymer science in this country." At the dedication, Paul Flory spoke on rubber elasticity. Arthur Bueche, vice president at General Electric and the brother of Kelley's Ph.D. advisor, gave the convocation address. Arthur M. Bueche to Maurice Morton, 1 August 1975, College of Polymer Science and Polymer Engineering Archives. Maurice Morton, undated form letter to invitees for the Whitby Hall celebration, College of Polymer Science and Polymer Engineering Archives. "Symposia, Paul Flory Talk, to Dedicate Akron U's New Whitby Hall," *Natural Rubber News* (September 1975): 3.

84. Frank Kelley, interview with the author, September 21, 2006.

85. William Muse, interview with the author, September 7, 2006.

86. Frank Kelley, interview with the author, March 16, 2006.

87. William Muse, interview with the author, September 7, 2006.

88. Ibid.

89. "AU Changes Polymer Research Site," *Akron Beacon Journal*, November 29, 1984.

90. Sharon Coleman, "AU Changes Polymer Research Site," *Akron Beacon Journal*, November 29, 1984.

91. John Kostrzewa, "Designing of Polymer Building Funded," *Akron Beacon Journal*, September 24, 1985.

92. William Muse, interview with the author, September 7, 2006.

93. Steven Litt, "Cleveland Architectural Firm Erecting New Goals," *Cleveland Plain Dealer*, February 16, 1997.

94. Fran Henry, "Richard Fleischman," *Cleveland Plain Dealer*, July 2, 1995.

95. Frank Kelley, interview with the author, September 21, 2006.

96. Diane Paparone Evans, "Architects Envision Glass Tower for Akron U's Polymer Center," *Akron Beacon Journal*, December 19, 1985.

97. Richard Fleischman, *Spaces to Be Shared* (Italy: l"Arca Edizioni, 1996).

98. Fleischman was not the sole designer of the building. The state of Ohio required that three separate architects work on the project. The other two firms were Harris/Day Architects of Massillon and Lawrence, Dykes, Bower and Clancy of Canton. The Lawrence firm had an expertise with laboratory facilities and also had a great deal of experience with the university, having designed its Bierce Library. "Polymer Building Architect Named," *Akron Beacon Journal*, November 20, 1986.

99. Susan Smith, "Nice Idea That Won't Fly: Plastic Building at AU," *Akron Beacon Journal*, November 12, 1987.

100. Susan Smith, "A Proposal That Was Way Ahead of Its Time," *Akron Beacon Journal*, November 12, 1987.

101. Susan Smith, "Nice Idea That Won't Fly: Plastic Building at AU," *Akron Beacon Journal*, November 12, 1987.

102. "University of Akron Begins Construction of $17 Million Polymer Science Building," 15 August 1988, University of Akron News Release, College of Polymer Science and Polymer Engineering Archives.

103. Andrea Louie, "AU Building Topped Off," *Akron Beacon Journal*, January 20, 1989.

104. For information on the budget specifics, see Donald L. Bowles to Frank Kelley, "Polymer Science Building Revised Budget," 21 March 1989, College of Polymer Science and Polymer Engineering Archives.

105. "Building Highlights," *Akron Polymer* 1, vol. 1 (Spring 1990): 7.

106. "Twenty-First Century Dawns at UA," Building Dedication Binder, College of Polymer Science and Polymer Engineering Archives.

107. Stephen Franklin, "As Big Wheels Exit, Akron Eyes New Treads," *Chicago Tribune*, July 28, 1991.

108. Rich Harris, "Tire Making Is Dying Out in Akron," *San Francisco Chronicle*, July 5, 1991.

109. "Ellen T. Otto to Frank Kelley, 15 April 1991, Building Dedication Binder, College of Polymer Science and Polymer Engineering Archives.

110. Cristal Williams Walker, "High-Tech Font of Knowledge Shines on Akron Skyline, Workplace at UA Deemed 'Ultimate,'" *Akron Beacon Journal*, December 3, 1990.

111. Cristal Williams Walker, "Putting on the Finishing Touches, UA's Polymer Building Weeks Away from Dedication," *Akron Beacon Journal*, March 15, 1991.

112. Letter from George V. Voinovich to attendees of the dedication ceremonies, 1 April 1991, College of Polymer Science and Polymer Engineering Archives.

113. "Tom Sawyer, quoted in "Here's What Others Are Saying about the Significance of the Polymer Science Building," Building Dedication Binder, College of Polymer Science and Polymer Engineering Archives. Before the building was occupied, a longtime member of the Department of Polymer Science retired. Barbara Zimmerman had initially joined the school in 1964 when she went to work for Maurice Morton. She recalled that he was an "exacting taskmaster." She went on to serve as Frank Kelley's secretary and in 1989 edited *Vignettes from the International Rubber Science Hall of Fame*. Taking over her duties was Rosemary Kolton, who would serve as Kelley's assistant during the move into the new polymer building and for a total of sixteen years. "Zimmerman Plans to Retire," *Akron Polymer* 1, vol. 2 (Fall 1990): 13.

114. Marketing material, College of Polymer Science and Polymer Engineering Archives. The university held dedication ceremonies for the new building on April 1, 1991. Though he was out of the country for the event, Robert Mercer from Goodyear said that the dedication would be "an occasion of international importance." A distinguished list of panelists and invited guests celebrated its opening over the course of several days. Sir Geoffrey Allen of Kobe Steel in London delivered the keynote speech, while John Lyons, the director of the National Institute of Science and Technology, presented one of the academic talks. Allen said he was "impressed by the novelty and obvious usefulness" of the new building. His presentation highlighted the importance of polymers to the public. The event also celebrated the achievements of Maurice Morton. Describing Morton as "one of the eminent architects" of the polymer program, the university awarded him an honorary doctorate. "Remarks Honoring Dr. Maurice Morton," Building Dedication Binder, College of Polymer Science and Polymer Engineering Archives.

115. Thrity Umrigar, "Instrument of Change, Some Criticize, Others Praise Departing Chief of University," *Akron Beacon Journal*, December 22, 1991.

116. "The Future of Polymer Research at the University of Akron," brochure, circa late 1980s, College of Polymer Science and Polymer Engineering Archives.

117. "Twenty-First Century Dawns at UA," Building Dedication Binder, College of Polymer Science and Polymer Engineering Archives.

118. Frank Kelley, "Challenges for a Rust-Belt Phoenix," 16 December 1993, Akron Roundtable presentation, College of Polymer Science and Polymer Engineering Archives.

CHAPTER 6

1. Joseph L. Nicholson and George R. Leighton, "Plastics Come of Age," *Harper's* (August 1942).

2. Frank Kelley, quoted in Webb Shaw, "From Old Warehouse to 3-Floor Miniplant," *Akron Beacon Journal*, May 20, 1985.

3. Eric Baer, quoted in "Polymer Pioneer," *Akron Polymer* 4, vol. 2 (Winter 1993): 6.

4. Sadhan C. Jana, interview with the author, March 16, 2006.

5. Edwin Layton, "Technology as Knowledge," *Technology and Culture* 15 (January 1974): 32.

6. Walter Vincenti, *What Engineers Know and How They Know It* (Baltimore, MD: Johns Hopkins University Press, 1990), 3–4.

7. Charles Slack, *Noble Obsession: Charles Goodyear, Thomas Hancock, and the Race to Unlock the Greatest Industrial Secret of the Nineteenth Century* (New York: Hyperion, 2002), 28.

8. Page Smith, *The Nation Comes of Age: A People's History of the Ante-Bellum Years* (New York: Penguin Books, 1981), 822.

9. E. Frankland, "Polymerization of Ethylic Cyanide," *Experimental Researches in Pure, Applied, and Physical Chemistry* (London: John VanVoorst, Paternoster Row, 1877), 59.

10. Alan Rocke, communications with the author, 10 October 2006; Alan Rocke, *The Quiet Revolution: Hermann Kolbe and the Science of Organic Chemistry* (Berkeley: University of California Press, 1993).

11. Jeffrey L. Meikle, *American Plastic: A Cultural History* (New Jersey: Rutgers University Press, 1995), 36; Baekeland, "The Syntheses, Constitution, and Uses of Bakelite," *Journal of Industrial and Engineering Chemistry* 1 (March 1909): 150–155.

12. Meikle, *American Plastic*, 10–11.

13. David Teegarden, *Polymer Chemistry: Introduction to an Indispensable Science* (Arlington, VA: National Science Teachers Association Press, 2004), 52–57.

14. Robert Friedel, "The Plastics Man," *Science* 5 (November 1984): 49.

15. *Time* 4 (22 September 1924); "Father of Plastics," *Time* 35 (20 May 1940): 50; Meikle, *American Plastic*, 33.

16. John Loadman, *Tears of the Tree: The Story of Rubber—A Modern Marvel* (Oxford: Oxford University Press, 2005), 195.

17. James White, interview with the author, October 24, 2006.

18. "Industrial Scholarships," *The Alumni Quarterly* (Spring 1917): 1.

19. "City University Aid to Industry," *Christian Science Monitor*, January 24, 1918.

20. Annual Report 1917, *The Bulletin of the Municipal University of Akron*, 18, University of Akron Archives.

21. George Knepper, *New Lamps for Old* (Akron, Ohio: University of Akron, 1970), 140–141.

22. Doctor of Philosophy in Polymer Science brochure, circa 1967, College of Polymer Science and Polymer Engineering Archives.

23. *Chemistry and Engineering News* (31 August 1964).

24. "Can Starch Aid Rubber Industry?" *Akron Beacon Journal*, July 17, 1966.

25. "AU Gains 2 New Research Contracts," *Akron Beacon Journal*, July 25, 1966.

26. Alan Gent to Michael Rzasa, 1 November 1967, College of Polymer Science and Polymer Engineering Archives.

27. Alan Gent to Serge Gratch, 23 October 1967, College of Polymer Science and Polymer Engineering Archives.

28. "A Plan for the '80s," The Institute of Polymer Science at the University of Akron, *Annual Report to the President*, 1979–80, University of Akron Archives.

29. Carl Snyder, quoted in "Polymer Valley/Vision or Fantasy for Akron," *Akron Beacon Journal*, June 12, 1983.

30. Sadhan Jana, interview with the author, March 16, 2006.

31. James White, interview with the author, October 24, 2006.

32. James White, in "Research Bright Spot for Area," *Akron Beacon Journal*, January 6, 1985.

33. James White, quoted in "Akron U Prepares for a Polymer Future," *Akron Beacon Journal*, April 28, 1983.

34. "Polymer Center Promised $1 Million in Gifts," *Akron Beacon Journal*, December 14, 1983.

35. Ibid.

36. James White, interview with the author, October 24, 2006.

37. Maureen Brown, "If He Ever Has an Ivory Tower, It Should Be Built of Polymers," *Akron Beacon Journal*, October 27, 1985.

38. "Rotational Molding Development Center Completes Fourth Year," *Akron Polymer* 1, vol.2 (Fall 1990): 5.

39. Lloyd Goettler, interview with the author, October 7, 2005.

40. Webb Shaw, "From Old Warehouse to 3-Floor Miniplant," *Akron Beacon Journal*, May 20, 1985.

41. "Remodeled Research Facility Unveiled in October," *Akron Polymer* 1, vol. 1 (Spring 1990): 8.

42. Frank Marini and Frank Kelley, quoted in "2 Akron U Deans to Retire in 1988," *Akron Beacon Journal*, May 20, 1987.

43. Nicholas Sylvester and Irving Miller, quoted in Diane Evans, "Dean's Mission Is Engineering Status Upgrade," *Akron Beacon Journal*, November 27, 1995.

44. Thein Kyu was born in Burma, where he studied physics at Rangoon University. He later earned his Ph.D. from the Kyoto Institute of Technology in textile engineering. Because of the material changes in textiles, this program changed its name and academic focus to become a department of polymer science and engineering. In 1982, he was a postdoctoral research fellow at McGill University in Canada and also a research associate at the University of Massachusetts. Late that year, he received a letter from James L. White, who was in Japan, inviting him to join White on the faculty in Akron. Kyu research included new ways to develop lightweight plastic materials for automobiles and aviation. Nancy Spinelli, "Thein Kyu," 1, 2 *Akron Polymer* 1, vol. 2 (Fall 1990): 10.

45. Avraam Isayev was the other founding faculty member in the polymer engineering program. Born in Azerbaijan, he received his master's of science degrees from Azerbaijan Institute of Oil and Chemistry in Chemical Engineering in 1964 and the Moscow Institute of Electronic Machine Building in Applied Mathematics in 1965. He then earned his Ph.D. at the Institute of Petrochemical Synthesis of the U.S.S.R. Academy of Sciences in 1970. Prior to his arrival at the University of Akron, he served as a senior research associate at Cornell University. He succeeded, and enjoyed building the polymer engineering program with White. "Dr. Avraam I. Isayev Finds Versatility a Plus," *Akron Polymer* 3, vol. 1 (Spring 1992): 10.

46. Cakmak was a chemical engineering graduate from the Technical University of Istanbul. In 1988, he earned a Presidential Young Investigator award from the National Science Foundation, a $500,000 grant over five years. He became the first professor in the polymer college to earn this honor, and only one of seven others in Ohio that year. His work on the structures,

properties, and processability of polymers, has attracted numerous sponsors, including Amoco, ALCOA, BF Goodrich, General Electric, and Eastman Kodak. "Mukerrem Cakmak," *Akron Polymer* 1, vol. 1 (Spring 1990): 12.

47. Min was tremendously enthusiastic about the opportunities offered by her profession. Her only complaint is the low number of women in polymer engineering. She said, "That needs to be changed, possibly by sending female representatives to colleges to recruit women for the polymer field." Kay Hehr, "Kyonsku Min-Cakmak," *Akron Polymer* 2, vol. 2 (Fall 1991): 9.

48. Han graduated with his bachelor's degree from Seoul National University and later earned his master's and doctoral degrees from the Massachusetts Institute of Technology. He began his career in Brooklyn in 1967, directing its polymer program starting in 1982. His work focuses on polymer theology, polymer physics, and polymer processing. Ong, quoted in "UA Names B.F. Goodrich Professor, Dr. Chang Dae Han of Brooklyn Is the First in Polymer Engineering," *Akron Beacon Journal*, August 24, 1992.

49. Stuart Drown, "Polymer Engineers Around the World Will Return to Akron for Conference," *Akron Beacon Journal*, April 3, 1994.

50. Tom Giffen, "Foreign Students Dominate UA Polymer Programs, Americans Are Clearly in Minority in Postgraduate Studies, University Attempting to Even Out Numbers," *Akron Beacon Journal*, May 26, 1992.

51. White, quoted in Katie Byard, "Necessity or Nuisance? Foreigners Abound in Graduate Schools," *Akron Beacon Journal*, March 26, 1990.

52. Other keys positions held by University of Akron polymer engineering alumni include project engineer at GE Plastics; senior staff chemist at Freudenberg in the United States; executive director of SAENG Thai Rubber Factory in Thailand; director of research and development at Mitsui Chemicals in Japan; senior technical development engineer at PolyOne in the United States; professor at Dankook University in South Korea; vice president and general manager at Abbott Labs in the United States; president of Orbital Plastics Consulting in the United States; assistant professor at the Suranaree University of Technology in Thailand; research specialist at 3M in the United States; assistant professor of materials engineering at Renssalaer Polytechnic Institute in the United States; senior process engineer at Johnson and Johnson Vision Care in the United States; and associate professor at the National University of Colombia.

53. Teresa Dixon Murray and Donna J. Robb, "Goodyear Seeks Aid," *Cleveland Plain Dealer*, March 6, 1999.

54. "EPIC-Hanna Blending and Compounding Center," *Akron Polymer* 1, vol. 1(Spring 1990): 5.

55. Douglas J. McGregor and James White, quoted in Katie Byard, "Akron U to Study Polymer 'Blending'," *Akron Beacon Journal*, October 24, 1989.

56. "Gifts Establish Mechanical Testing Lab," *Akron Polymer* 2, vol. 1 (Spring 1991): 3.

57. "UA Engineers Develop Program for Co-Rotating Twin Screw Extruder," *Akron Polymer* 3, vol. 2 (Fall 1992): 3.

58. Kielar, quoted in Rebecca Yerak, "Putting Dreams to Work School Finds Uses for Basic Research in Polymers," *Plain Dealer*, September 13, 1992.

59. Lornet Turnbull, Jim Quinn, and Yalinda Rhoden, "City Is Poised for Takeoff, But Where Will It Go?" *Akron Beacon Journal*, July 26, 1992.

60. Stuart Drown, "Polymers—They're Just Everywhere," *Akron Beacon Journal*, July 20, 1994.

61. Glenn Gamboa, "AES Off and Running," *Akron Beacon Journal*, June 9, 1996.

62. Ginter, quoted in Lornet Turnbull, *Akron Beacon Journal*, July 21, 1994.

63. Mary Ethridge, "Praises Abound for AES Builders," *Akron Beacon Journal*, February 24, 1996.

64. Steve Love and David Giffels, *Wheels of Fortune* (Akron, OH: University of Akron Press, 1999), 295–296.

65. Ken Keuchel, quoted in Janet Moore, "Success Sticks to Entrepreneurs, Engineers Quickly Developing New Markets," *Akron Beacon Journal*, February 27, 1996.

66. Mary Vanac, "Inter-Ion Makes Mark in Plastic," *Akron Beacon Journal*, September 29, 1994. .

67. John Russell, "Area Evolves with Polymers," *Akron Beacon Journal*, December 27, 1998.

68. Frank Kelley, "Dean's Message," *Akron Polymer* 7, vol. 1 (Fall 1996): 2.

69. Frank Kelley, interview with the author, September 21, 2006.

70. Marshall Dean, quoted in "Donation Stretches Polymer Research," *Cleveland Plain Dealer*, October 22, 1997.

71. Luis M. Proenza, quoted in "Polymer Engineering Academic Center," *Akron Polymer* (Spring 2001): 2.

72. Katie Byard, "University Digs into $200 Makeover," *Akron Beacon Journal*, December 1, 1999.

73. Sarah Thorley, "New Home for Polymer Engineering," *Connect with DPE* (2005): 2.

74. Katie Byard, "Future Polymer Engineers to Get New UA Home," *Akron Beacon Journal*, January 16, 2001.

75. " 'Mission Accomplished' for Dr. James L. White," *Akron Polymer* (1999): 5.

76. Sadhan Jana, interview with the author, March 16, 2006.

77. Katie Byard, "Polymer Expert to Lead Program, UA Hires R. Byron Pipes as Head of New Akron Global Polymer Academy," *Akron Beacon Journal*, September 26, 2001.

78. Justin Molenaur, "Akron Global Polymer Academy," *Connect with DPE* (2005): 2.

79. R. Byron Pipes, quoted in Stephen K. Ritter, "A Lifetime of Polymer Learning," *Chemical & Engineering News* 80, vol. 4 (28 January 2002): 69–71.

80. Kevin Cavicchi, interview with Justin Molenaur, May 30, 2006.

81. Mark D. Soucek, interview with Justin Molenaur, May 30, 2006.

82. "Sutton Announces Federal Grant for University of Akron," July 12, 2007, *http://sutton.house.gov/070713grant.htm*.

83. Lloyd Goettler, interview Justin Molenaur, May 30, 2006.

CHAPTER 7

1. Adam Rogers, et. al. "A New Brand of Tech Cities," *Newsweek* 137, vol. 18 (April 30, 2001): 44–51.

2. Donald Plusquellic, quoted in Betty Lin-Fisher, "Akron 'New Tech City,' *Newsweek* Names 10 That Have Built New Economies," *Akron Beacon Journal*, April 25, 2001.

3. John Russell, "The Proenza Push," *Akron Beacon Journal*, March 12, 2002.

4. These included a Summer Institute in Polymer Science for College Chemistry Teachers and an undergraduate research program.

5. Morton, quoted in "Dr. Maurice Morton: Asking the Right Questions, and Accepting no Compromises," *Akron Polymer* 4, vol. 1 (Spring 1993): 7.

6. Ron Kirksey, "Weighing Peggy Gordon Elliot's Impact," *Akron Beacon Journal*, August 27, 1995.

7. Peggy Gordon Elliot, *The Urban Campus: Educating the New Majority for the New Century* (Phoeniz, AZ: The Oryx Press, 1994).

8. Abe Zaidan, "Elliot's Future in Doubt," *Cleveland Plain Dealer*, March 29, 1996.

9. Roger J. Metzger and Ron Kirksey, "Elliot Won't Go from UA on Own," *Akron Beacon Journal*, March 31, 1996.

10. Advancement Council Letter, in "More Correspondence on Governance of the University of Akron," *Akron Beacon Journal*, February 14, 1996.

11. Luis M. Proenza, interview with the author, November 1, 2006.

12. "Proenza Biography," at *http://www3.uakron.edu/home/proenza.html*.

13. Luis M. Proenza, interview with the author, November 1, 2006.

14. Katie Byard, "Purdue Administrator to Lead UA," *Akron Beacon Journal*, October 15, 1998.

15. Luis M. Proenza, interview with the author, November 1, 2006.

16. Proenza, quoted in Katie Byard, "A New Era at the University of Akron, The Proenza Plan," *Akron Beacon Journal*, October 10, 1999.

17. Luis M. Proenza, "Ohio Must Spend More on Research and Development," *Cleveland Plain Dealer*, July 20, 1999.

18. Ibid.

19. Proenza quoted in, John Russell, "The Proenza Push," *Akron Beacon Journal*, March 12, 2002.

20. Ibid.

21. Kelley, quoted in Abe Zaidan, "Ohio Warned of Clouds on Horizon of its Success in Plastics Industry," *Cleveland Plain Dealer*, December 17, 1993.

22. Kelley, quoted in Abe Zaidan, "Polymer Scholars Bring the World to UA Symposium," *Cleveland Plain Dealer*, July 11, 2004.

23. Kennedy, quoted in Stuart Drown, "Akron Will Host High-Tech Talk, 1700 Polymer Researcher from Around the World Coming to Town to Discuss, Learn More about Polymers," *Akron Beacon Journal*, July 8, 1994; Stuart Drown, "MacroAkron's Driving Force," *Akron Beacon Journal*, July 10, 1984.

24. Abe Zaidan, "International Scientists Visit 'Polymer Valley,' " *Cleveland Plain Dealer*, July 12, 1994.

25. Stuart Drown, "Ohio Industry Fights to Keep Plastics Edge, Meeting Will Address State's Danger of Losing Lead in $15 Billion Trade," *Akron Beacon Journal*, April 3, 1994.

26. Russell Livigni, quoted in "GenCorp Foundation Gives UA First Check," *Akron Beacon Journal*, January 11, 1996.

27. Purcell, quoted in Rebecca Yerak, "Putting Dreams to Work School Finds Uses for Basic Research in Polymers," *Cleveland Plain Dealer*, September 13, 1992.

28. John Russell, "EPIC's Proportions Growing," *Akron Beacon Journal*, October 7, 1998.

29. "PVC Seeks Breakthrough Improvements," *Akron Polymer* (Spring 2000): 10.

30. Katie Byard, "UA Opens a New Pipeline to PVC Research," *Akron Beacon Journal*, October 29, 1998.

31. Frank Kelley, interview with the author, September 21, 2006.

32. Lando and Gibara, quoted in John Russell, "Flexible Ideas, But Brittle Results," *Akron Beacon Journal*, March 10, 2002.

33. Frank Kelley, interview with the author, September 21, 2006.

34. Pat Valente and William Overton, quoted in John Russell, "Polymer Group Loses Funds," *Akron Beacon Journal*, October 12, 2000.

35. Charles Wilkes, quoted in "Group Molding Companies Non-Profit Organization Looks to Help Turn Ideas Into Polymer Businesses, Products," *Akron Beacon Journal*, July 20, 2001.

36. Charles Wilkes, interview with the author, September 8, 2006.

37. John Russell, "Polymers Are Getting a Big Push," *Akron Beacon Journal*, October 5, 2001.

38. Frank Harris, quoted in "Dr. Frank Harris Juggles a Hefty Variety of Responsibilities," *Akron Polymer* 4, vol. 2 (Winter 1993): 9.

39. Peter Krouse, "Promoting the Promise, Prominence of Polymers," *Cleveland Plain Dealer*, July 28, 2002.

40. Ron Clark, quoted in John Russell, "Akron Inches Along Toward Polymer Goals," *Akron Beacon Journal*, November 10, 2002.

41. John Russell, "Ohio Pulls Plug on Money for Local Effort to Establish Akron as a Global Center," *Akron Beacon Journal*, April 8, 2003.

42. Frank Kelley, interview with the author, September 21, 2006.

43. Editorial, "Detour to Polymer Valley—The State Takes a Wrong Turn," *Akron Beacon Journal*, April 9, 2003.

44. John Russell, "OPED's Demise a Game of Blame," *Akron Beacon Journal*, August 17, 2003.

45. Charles Wilkes, interview with the author, September 8, 2006.

46. Luis Proenza, interview with the author, November 1, 2006.

47. PolymerOhio, at *http://www.polymerohio.org/Home/tabid/36/Default.aspx*.

48. Newkome, quoted in John Russell, "The Proenza Push," *Akron Beacon Journal*, March 12, 2002.

49. Ohio Department of Development, "Third Frontier Project and Related Programs," at *http://www.odod.state.oh.us/tech/program.htm*.

50. Pat Valente, quoted in John Russell, "State Denies University's Bid for $10 Million to Build Polymer Center," *Akron Beacon Journal*, June 5, 2003.

51. Evaluation Committee, as quoted in John Russell, "UA Loses Out Again in Bid for State Funds," *Akron Beacon Journal*, November 4, 2003.

52. John Russell, "State to Trim Plan for Economy," *Akron Beacon Journal*, November 6, 2003.

53. John Russell, "UA Revamps Pitch for Aid from State," *Akron Beacon Journal*, August 20, 2004.

54. Dave Scott, "Polymers Get Boost from State," *Akron Beacon Journal*, May 11, 2005.

55. Dave Scott, "Polymer Funding to Yield Products," *Akron Beacon Journal*, May 14, 2005.

56. Sharell Mikesell, "State Announces $34 Million to Support Polymer and Advanced Materials Industries," at *http://www.polymerohio.org/Portals/0/PressReleases/OHPolymerWCIRelease.pdf*.

57. Luis Proenza, "State Announces $34 Million to Support Polymer and Advanced Materials Industries," at *http://www.polymerohio.org/Portals/0/PressReleases/OHPolymerWCIRelease.pdf.*

58. Mukerrem Cakmak, interview with Justin Molenaur, May 30, 2006.

59. "Layering Polymers," *Akron Beacon Journal*, July 30, 2006.

60. Luis Proenza, "Regional Imperatives in Northeast Ohio," National Summit on Regional Innovation, 22 April 2005, at *http://www.compete.org/pdf/Luis_Proenza_University%20of%20Akroni.pdf.*

61. "Business & Lobbying," *The Hill* (September 5, 2006).

62. Peter Krouse, "Grants Fuel Polymer Research," *Cleveland Plain Dealer*, May 14, 2005.

63. Luis Proenza, interview with the author, November 1, 2006.

64. John Russell, "The Proenza Push," *Akron Beacon Journal*, March 12, 2002.

65. Denise Grady, "Gravity-Defying Geckos Teach Scientists a Lesson," *New York Times*, August 30, 2005.

66. "Leaping Lizards! Team at UA Makes Gecko Tape," *Akron Beacon Journal*, July 1, 2007.

67. Thomas J. Sheeran, "Pentagon Is Looking to Nature for Tougher but Lighter Materials," *Philadelphia Inquirer*, February 21, 1993.

68. These include the Ticona Award for Excellence in Polymer Science ($500 awarded for best presentation of a formal seminar); the Rubber & Plastics News Award ($500 awarded for best performance in a research presentation in polymer science); the Maurice Morton Award ($1,000 awarded for highest GPA in core polymer science courses during the first year of study); the Ronald K. Eby Award ($1,000 awarded for highest GPA in core polymer science courses during the first year of study); the Omnova Solutions Signature Award ($2,500 awarded to a student in the third year of study for best cumulative exam performance); the Hoyt F. Wells Scholarship ($5,000 awarded to a student who has performed with excellence in coursework, exams, and research during the third year of study); the Eastman Chemical Company Enhancement Scholarship ($5,000 awarded to top third-year student from Akron selected by Eastman); and the Goodyear Enhancement Fellowship ($7,500 awarded to top U.S. or permanent resident student in the fourth year of study). "Outstanding Polymer Science Students Are Honored at May 2 Event," at *http://www.uakron.edu/news/articles/uamain_467.php.*

69. "James E. McGrath Builds Solid Team at Virginia Tech," *Akron Polymer* 2, vol. 1 (Spring 1991): 2.

70. "Nissim Calderon," *Akron Polymer* 1, vol. 1 (Spring 1990): 3.

71. "Russell A. Livigni," *Akron Polymer* 1, vol. 2 (Fall 1990): 3.

72. "Dr. Sharell Mikesell," *Akron Polymer* 3, vol. 2 (Fall 1992): 5.

73. Diane Evans, "Dean Links Business, Polymers Grad Students to Study Specialized Science," *Akron Beacon Journal*, June 27, 1999.

74. Mark Foster, interview with the author, October 7, 2005.

75. Katie Byard, "Necessity or Nuisance? Foreigners Abound in Graduate Schools," *Akron Beacon Journal*, March 26, 1990.

76. "Linking Resources for Science," *Akron Beacon Journal*, April 24, 1992.

77. Julie Newhall, "Elementary Students Form a Link in the Polymer Chain," *Akron Polymer* 3, vol. 1 (Spring 1992): 3.

78. Clem, quoted in "Akron Polymer Training Center," *Plain Dealer*, October 4, 1994.

79. Frank Kelley, "Dean's Message," *Akron Polymer* (Fall/Winter 1998): 2.

80. Rod Quirk, quoted in Jim Mackinnon, "UA Polymer Program Turns 50," *Akron Beacon Journal* (8 August 2006).

81. Arden L. Bement, quoted in Jim Mackinnon, "National Science Foundation Director Speaks at UA," *Akron Beacon Journal*, August 12, 2006.

82. Ibid.

83. Mike Hixenbaugh, "Polymer Dean Leaves Nation's Top Program," *The Buchtelite*, August, 28, 2006.

84. Joseph P. Kennedy, interview with the author, June 20, 2006.

85. David Giffels, "Vision Shapes Akron's Future," *Akron Beacon Journal*, December 17, 2006.

86. Kelley, quoted in Rebecca Yerak, "Putting Dreams to Work School Finds Uses for Basic Research in Polymers," *Cleveland Plain Dealer*, September 13, 1992.

87. Dorothy Shinn, "New Akron Landmark Is Crystal Blue," *Akron Beacon Journal*, July 31, 2005.

88. Luis Proenza, quoted in Carol Biliczky, "UA's New Artistic Gem Shines Through Clouds," *Akron Beacon Journal*, November 10, 2005.

89. "Retiring Dean Has Admirable Legacy," *Akron* (Fall 2006): 11; additional interview excerpts at *http://www2.uakron.edu/cpspe/highlight.html*.

90. Stephen Cheng, email to author, July 20, 2007.

91. Kevin Burgess, "Organometallic Chemistry," *Chemistry & Industry* (February 21, 2000): 153.

92. Roger J. Mezger, "Polymer Valley Wins National Recognition," *Akron Beacon Journal*, March 21, 1996.

93. Proenza, quoted in John Russell, "The Proenza Push," *Akron Beacon Journal*, March, 12, 2002.

94. "The Polymer Difference," *Akron Beacon Journal*, August 13, 2006.

95. Gibara, quoted in "The Polymer Difference," *Akron Beacon Journal*, August 13, 2006.

APPENDICES

1. "Secret and Confidential, War Production Board," Minutes of Meeting #1 in Akron, 28–29 December 1942, College of Polymer Science and Polymer Engineering Archives.

2. Advertisement for "A Technical Course in Rubber Manufacture," *The Rubber Age* (25 June, 1917): 268–269.

3. "Educational Courses in Rubber Chemistry and Technology," *Rubber Red Book* (1955–56): 981–989.

4. Sadhan Jana, interview with the author, March 16, 2006.

5. Llody Goettler, quoted in "Lloyd A. Goettler Becomes PE Department Chair," *Akron Polymer* (Spring 2000): 13.

6. Sadhan Jana, quoted in "Polymer Engineering Professor Excited to Be in Akron," *Akron Polymer* (1998): 8.

7. Mark Foster, interview with the author, October 7, 2005.

8. Donald McIntyre, quoted in "Donald McIntyre Focuses on 'Unity, Liberty, and Charity," *Akron Polymer* 4, vol. 1 (Spring 1993): 11.

9. "Quirk," *Akron Polymer* 2, vol. 2 (Fall 1991): 9.

10. "Quirk Becomes Chair of Polymer Science Department," *Akron Polymer* 6, vol. 1 (Fall 1995): 6.

11. William Brittain, quoted in "Dr. William Brittain Involved in Creating New Polymers," *Akron Polymer* 5, vol. 1 (September 1994): 9.

12. "Stephen Z.D. Cheng," *Akron Polymer* 1, vol. 1 (Spring 1990): 13.

13. "Foster Champions Science," *Akron Polymer* 6, vol. 1 (Fall 1995): 10.

14. Mark Foster, interview with the author, October 7, 2005.

15. "New Institute Director Named," *Akron Polymer* 1, vol. 1 (Spring 1990): 9.

16. "Frank Harris Named Outstanding Researcher," *Akron Polymer* 3, vol. 2 (Fall 1992): 7.

17. "Polymer Director Is Chosen at University, Professor Frank Harris to Oversee UA's Maurice Morton Institute," *Akron Beacon Journal*, October 7, 1994.

18. "Ohio Polymer Strategy Council," at *http://www.polymerohio.org/PolymerOhio/OhioPolymerStrategyCouncil/tabid/106/Default.aspx.*

INDEX

ABOUT THE AUTHOR

Mark D. Bowles has authored or coauthored seven books on the history of space exploration, nuclear research, and medicine. His book, *Science in Flux*, won the American Institute of Aeronautics and Astronautics' 2005 History Manuscript Award, which is presented each year for the best historical manuscript dealing with the impact of aeronautics and astronautics on society. He received his B.A. in Psychology (1991) and M.A. in History (1993) from the University of Akron. He earned his Ph.D. in the History of Technology (1999) from Case Western Reserve University and his MBA in Technology Management from the University of Phoenix (2005). He founded Belle History in 2007. He is married to Nancy and they have a daughter, Isabelle. He can be reached at mark@bellehistory.com.